STEEL RESOLVE

STEEL RESOLVE

A STORY OF ENDURANCE

Hugh W. Williams II

STEEL PUBLISHING

Omaha, Nebraska

ISBN10 - 0-9820589-0-X
ISBN13 - 978-0-9820589-0-9
Library of Congress Control Number: 2008935367
Cataloging in Publication Data on file with publisher.

Drake-Williams Steel Press
2301 Hickory Street
Omaha, Nebraska 68108
www.DWSteel.com

Book Design: Gary James Withrow
Production, Distribution, and Marketing: Concierge Marketing Inc.

Printed in the United States of America
10 9 8 7 6 5 4 3 2 1

CONTENTS

PREFACE

It is an amazing story of survival to have a company endure for more than 125 years under the continuing direction of one family.

It would be amazing if that were true. But, of course, it isn't.

This is the story largely of a company dating to 1882 that has morphed into different forms fully five times. Not total changes, but at least name changes for sure. Let's face it—most firms go broke, are sold, or simply are dissolved over that much time.

Our company has existed through two world wars plus other wars and events not as widespread, but plenty serious; depressions, recessions, money panics, and flu epidemics.

Our company has been like one of our meandering Midwestern rivers, wending along kind of slowly, and taking the path of least resistance. It has had instances of stagnation, where it has almost dried up. It has had to be dredged and cleaned up so it could flow freely again. It has come near to dying.

I have researched the facts as well as I could. Some of the myths had to be exploded:

1. Hugh Williams was *not* one of the founders of the company.

2. Andrew Carnegie probably never even heard of the company.

3. *Mount* was a real person, and was *not* an activity of the firm (mounting?).

You will read who the real founder was, and how little we know of him. The Williams family has been associated with the company, probably since 1897, and in charge continuously since the early 1900s.

I hope you enjoy the whole story.

Hugh W. Williams, II

FOREWORD

THE HISTORY AND PROGRESS OF THIS COMPANY IS MIND-BOGGLING. We started in the horse and buggy era. Now we have high-tech, computer-controlled equipment that I'm sure would dazzle our employees of the past.

I started working at Drake-Williams-Mount Co. in the spring of 1959. I was assigned to be a helper on the quarter-inch shear and the old leaf brake with a man named Tony McAndrews. He always called me "Kid." We worked in the old East Shop. Our area had no overhead cranes or very good lighting. It had a dirt floor, broken-out windows, and a roof that leaked when it rained. The only way to move materials was on an old cart that was mounted on a simple rail system in the dirt floor. The yard wasn't paved and was full of potholes. We had twenty-eight employees and our sales volume wouldn't cover much more than a couple of salaries today.

Today, Drake-Williams Steel is a multi-million-dollar enterprise, employing over 200 people and producing more than 30,000 tons of structural steel fabrication and another 30,000 tons of rebar. We have three plant locations in the Omaha area and sales offices in Denver and Des Moines. We have really been fortunate to help shape the skyline of Omaha and many other places with beautifully built, long-lasting structures. When I drive around, I'm filled with memories … and pride.

To have been a part of this story line, from what was really a ramshackle dump to today's strong, successful company, has been fun and very satisfying. We owe it all to the many good, dedicated employees who've worked here all their lives and would live and die for the company. Thank you, to them. We all share this milestone.

The future of the company is in very good hands, and only good things are coming. I hope you enjoy this historical account. Long live Drake-Williams Steel!

Michael C. Williams

INTRODUCTION

As I have done this research, it has been interesting and fascinating, really, to see the parallels—that of our company with the history of both Omaha and the nation. It gives a certain perspective and better understanding of the actions and reactions by company leaders along the way.

That a company could survive and prosper, small and closely held! The old adage goes, "the first generation sows it, the second grows it, and the third blows it," or something like that. Sometimes we Williamses clung on for dear life, and at other times we stepped out and took big risks. But never, never did we "milk" the company of assets, and it was always run for the benefit of all of its employees.

It is a shame more is not known of the early history of this company. A great deal transpired, and our fortunes were tied in so closely with the fortunes of Omaha and this region. A small company is not easy to operate—the time demands are great on the managers and principals.

Keeping the wheels turning, fending off competition, the agencies of government, coping with demands of unions, customers, suppliers, bankers—really, a full plate. Scarce is the time, if any, remaining for keeping a diary or journal of daily events. Company records are thrown out. The sheer volume if retained over all the years would be overwhelming. Photographs are wonderful, but all too few. Some are lost. Memories fade, key people die and are replaced. But the wheels keep turning.

This is the story of a company, but a company, of course, is its people. Land, buildings, equipment, rolling stock, inventories—all are possessed by different companies and are essentially the same. But people make the difference for success or failure. We tend to think about the former more, the latter less. And yet, the most valuable assets are those people. And frighteningly, they all go home every night to their homes and families. We trust that they will return the next day, bringing with them their invaluable skills and talents, plus the experience and essential mental "records" that they carry with them!

Every generation stands squarely on the shoulders of the previous one. A great deal of gratitude is due all of the "ones that came before," the ones that persevered, that hung on against all of the odds, and some who even had visions of bigger and better tomorrows.

So this story is about them, and we owe them much. I am very grateful, and proud, too, to have been a part of this tradition and tremendous growth—literally from the dirt floors of our former shops to what we have today: A modern, thriving company, with many structures, great facilities and equipment, but above all, many great people.

This book, then, is dedicated to them—to all the men and women who came before!

ONE

EARLY HISTORY: THE BEGINNINGS

The story goes that Hugh Williams (the original) was journeying back to Wales with his father, William, from whence William had emigrated in his youth. Along the way and aboard ship, he had the good fortune to meet Andrew Carnegie, and apparently made friends with the famous industrialist and later equally famous philanthropist.

Sometime later, evidently about 1881, and after Hugh had made the move from Chicago to Omaha, he was stopped on the board sidewalk of downtown Omaha by a Western Union messenger boy who had been trying to find a "Hugh Williams." At the office of Western Union was a message and also a money order for $3,000, an outright gift from Carnegie. The message informed him of a blacksmith shop in Sioux City that had gone bankrupt and suggested he go there and buy the equipment. He did so and moved the equipment to Omaha on horse-drawn drayage wagons, setting up the first shop at 19th and Pierce streets.

Now, there is a real tale! One with romance, intrigue, and singular good fortune to it. That story was told to me by my cousin Georgia Lee, daughter of Alice Benson who was my father Arthur's older sister. Georgia swears that the story is true, at least as told to her by her mother, Alice. Just imagine, the great Andrew Carnegie, the man who built a steel empire and endowed 2,509 libraries in towns across this country. During his lifetime, it is estimated that he gave away over $350 million. Why, we could use this story and tell and retell it without fear of contradiction, and no one the wiser. What a way to found a business!

In doing a little superficial research on the life of Andrew Carnegie, I found that he did escort his mother on a trip to Dunfermline, Scotland, in 1881. Could it be?

But the truth seems to lie in other directions and involves much more mundane happenings, plus quite a lot of risk taking. Here, then, is the story, as near as I can piece it together, and, believe me, much of the very early history is not documented. Some comes from small newspaper items, some from correspondence, some from city and county directories, and finally, some from anecdotal evidence passed along through the generations.

This is a company that has seen six separate name changes over the years. Here is the chronology:

1882- founded as the Wilson Steam Boiler Co.

1883- Wilson Boiler and Sheet Iron Works

1893- Mid-Continent Boiler and Sheet Iron Works,
Wilson & Drake, Proprietors

1897- Drake, Wilson and Williams

1902- Drake-Williams-Mount Co. (first incorporation)

1958- Drake-Williams Steel, Inc.

The year 1882 was quite a year, in that this was the year that the rotating magnetic field was discovered by Nikola Tesla. This discovery was fundamental to the invention by Tesla also, of the A.C. induction motor, and the generation and transmission of electrical energy over long distances. Prior to this, motors were D.C., as A.C. was not considered safe nor even possible.

Today, the unit of magnetism or of magnetic resonance is called the "Tesla." Tesla invented many, many items, and when he died in 1943, he held over 700 patents in the United States and Europe. It is sad that he is not given more notice in our history books and more credit for all that he accomplished.

The original name, then, was Wilson Steam Boiler Co., and no Williams family member had anything to do with this company. It was founded by John Morris Wilson in 1882, and very little is known about the early day operations. We know that he must have been a mechanical engineer since he designed and built complete steam boilers under the name "Wilson."

In the files at the Omaha Public Library for the Wilson Steam Boiler Co. is a very fragile page, with what must be the earliest pictures of the company in existence. One shows the brick factory building with "Wilson Steam Boiler Co." across the front. The second picture is of "Mr. J.M. Wilson" looking very Western, with handlebar mustache, vest and coat, dark wide-brimmed hat, and string tie. The third picture is of "Supt. and his men," with sixteen shop men in the picture, including Wilson at the rear of the group. Company offerings: Water Tubular and Internal Fire Boilers, Stacks and Breechings, Air Pressure Tanks, Steam Jacketed Tanks, Rendering Tanks, Grain Storage Tanks, Sheep

Dip Tanks, Oil and Water Storage Tanks, Gas Holders.

In the description, it was said to be established in 1882 at the corner of 19th and Pierce streets and incorporated in 1902, with J.M. Wilson as President and J. Morris Wilson secretary and treasurer. It would appear that John Morris Wilson was putting "Mr. J.M. Wilson" in the presidency and himself as secretary-treasurer, which is the only reference anywhere to a junior Wilson. The probability is that "J.M. Wilson" is the son of John Morris. If this account is accurate, it does not follow that the name Wilson is dropped, as the firm becomes Drake-Williams-Mount Co. in 1902. Evidently, the Wilson interests were bought out in that year.

From the reference book, *Leading Industries of the West: Omaha*, is the entry "Wilson Boiler Works," established 1883 (elsewhere, the date has always been 1882) with a lot measuring 50 x 144 feet, and building 40 x 85 feet. Further, that the company has "12 expert mechanics," and that "Wilson learned his trade in New York City, and has 42 years experience in the business."

John Morris Wilson, founder of Wilson Steam Boiler Works, forerunner of Drake-Williams Steel, Inc.

We have one picture of Wilson, taken from the book *Nebraskans 1854–1904* in which he is simply referred to as "Proprietor Wilson Steam Boiler Co., Omaha." He is strong looking, with a bushy mustache, close-cropped hair, and a firm looking jaw. In the City Directory of 1883-84, Wilson was listed as residing at 2022 St. Marys Ave. In the City Directory of 1893, some ten years later, he is shown as living at 641 So. 17th St.

Early "Big Doings" in Omaha

This company, which was just being formed, could not possibly have attracted too much notice, and certainly not the kind of publicity as another event of the day. Here is what was going on around the same time. It was an event that had been hyped in Omaha's newspapers for weeks in advance of the big day, May 19, 1883:

THE SUNSHINE OF SUCCESS
At last, illuminates the Great Garden
Of Merit, where the Wild Flowers
Of Western Talent are destined to
Bloom in rare and extraordinary profusion
EVERYTHING INDICATES THAT THE EVE OF AN EVENT
Which will mark an epoch in the History of the
Amusement Universe has at length arrived, and
That Cody & Carver's
"WILD WEST" WILL BE POSITIVELY INAUGURATED
Saturday, May 19th, as Advertised

This was the notice carried in the *Omaha Bee* newspaper and was the first performance anywhere of Buffalo Bill and his Wild West show, which continued on well into the twentieth century, with thousands of performances. It became an institution, and there is more than an even chance that Hugh and the others took in at least one performance.

But the first season was a turbulent one and lost money. They did a show in Council Bluffs; Springfield, Illinois; and Chicago, and on east. Dr. Carver was a heavy drinker and became jealous of Buffalo Bill's popularity. For Bill Cody, it was a year of marital strife, plus the loss of his eleven-year-old daughter. The show returned to Omaha, and they divided their assets by coin toss. Carver went on with his own Wild West show but never attained the popularity or acceptance of Buffalo Bill's Wild West show.

It is intriguing to realize that these men who were founding Wilson Steam Boiler were living contemporaneously with Buffalo Bill, General George Crook, and George Armstrong Custer. The famous disaster at the Little Big Horn was not much earlier, in 1876. The great Civil War was during Hugh's lifetime, and it was only about 1869 that the railroad was completed to the West Coast. And the first bridge across the Missouri (a spectacular engineering achievement for the time) was not completed until 1872.

Also, in 1879 a landmark case was tried in Omaha, and while it was called the trial of Standing Bear, an Indian, it was actually Standing Bear who put the United States government on trial.

Standing Bear and his people, the Ponca, had been banished from their tribal home in the Niobrara River valley to the Indian Territory, in Oklahoma—an action that was brought in violation of the treaty that had been signed and that should have provided the Ponca a permanent home in Dakota. Their plight came to the attention of Thomas Tibbles of the *Omaha Herald*. Standing Bear, after making the excruciatingly painful journey all the way back from Oklahoma, was on his way to Dakota to fulfill his promise to bury his son in Dakota. The son was just one of the many Ponca who had died from hunger and sickness in his tribe, and the rest who had no means of making a living in their new "home."

Tibbles became a hero in the defense of Standing Bear, as he decided to take on the United States in a court of law. Other attorneys were part of the legal team. You have driven on the Omaha streets many times that now bear the names of his attorneys: Poppleton (A.J.) and Webster (John L.). The trial was held before Judge Elmer Dundy in May. The task before them looked hopeless, the odds immense. Nobody, but nobody, sued the U.S. government.

As it was not a criminal case, there was no jury. *The New York Times* picked up the story, and the headline asked "Have Indians Any Rights?" The attorneys for Standing Bear argued that he was no longer a "Chief," and that he wanted to live as a white; therefore, he should be entitled to his rights under the Fourteenth Amendment.

However, Standing Bear had his own agenda. While the other Poncas appeared in white man's clothing, Standing Bear wore his full regalia as a Chief: headdress, bear claw necklace, and moccasins on his feet. However, he did testify that he wanted to work and become "like a white man." The attorneys argued and quibbled the technical points, and the government case was based in part on the Dred Scott decision by the Supreme Court, which had found essentially that a Negro slave did not have the same rights as a "person."

For Standing Bear, what mattered was this: "From the time I went down there (to Indian Territory) until I left, one-hundred and fifty-eight of us died. I thought to myself, God wants me to live, and I think if I come back to my old reservation, he will let me live. What have I done? I am brought here, but what have I done? I don't know."

As the trial ran down, Standing Bear wished to make a speech. Although this was not normally allowed, Judge Dundy let him proceed. In the best known scene as described by Thomas Tibbles:

Standing Bear rose and addressed the audience. He stretched his right hand out before him, holding it still so long, the audience grew tense.

At last, looking up at the judge, he spoke quietly. "That hand is not the color of yours, but if I pierce it, I shall feel pain. If you pierce your hand, you will also feel pain. The blood that will flow from mine will be the same color as yours. I am a man. You are a man. The same God has made us both."

It turns out that this speech, cloaked in great drama, was embellished by newsman Tibbles. But it is true that Standing Bear made a rather long speech in that courtroom, and he spoke of his efforts to learn to farm and live as a white man, and his words carried a ring of truth. The judge ruled in his favor, and so Standing Bear vs. Crook became a very important civil rights case and marked the first time that a U.S. court had ruled that an Indian was officially a person.

Forgive another diversion from our story to observe another notable event in Omaha in that remarkable year, 1882. It was in the fall that citizens were called upon to vote on a constitutional amendment granting women the right to vote. Edward Rosewater, the editor and publisher of the *Omaha Bee*, accepted the challenge of Susan B. Anthony, the famous suffragette, to a debate on the subject. This debate took place at Boyd's Opera House in September of that year. Anthony was quoted in the *Bee*, October 14, 1882, as follows:

Notwithstanding the high flown declarations in our declaration of independence our government was founded on the bones and blood of slavery and on the disenfranchisement of half its population.

Here is Rosewater, praising the mother of the late president, during his debate with Susan B. Anthony:

In my estimation, the highest aim that a woman in this country may have, more than she could attain by the enfranchisement of her sex—to be the mother of a manly man like James A. Garfield.

More Omaha Oddities

A little earlier, I mentioned St. Marys Avenue, and if you have ever wondered why this avenue runs at an angle, and not parallel to the other east-west streets, it is due to an early "mover and shaker" of note, Mr. Harrison Johnson, who lived in a log house (homestead) just west of Mr. Woolworth's present home (wherever that was), but west some distance from the downtown area. The street was built along a straight line, but which angled directly to the village of Omaha.

From the *History of the City of Omaha* (Savage and Bell, 1894, p. 494), we find under Manufacturing Interests, "The Mid-Continent Boiler Works, corner of Pierce and 19th St., Wilson and Drake Proprietors, carry on an extensive business in the line of boilers, tanks, etc., with $30,000 invested. The establishment employs thirty men."

It is interesting to note another firm described as follows: "Carter and Son have been engaged in the manufacture of boilers, tanks, smokestacks, etc. in Omaha since 1888, on the B & M railway tracks at the intersection of 20th St. Twenty men are employed by this firm." This possibly puts this outfit a few blocks south, probably about Woolworth Street on the east side of 20th Street, because we know the U.S. Army was on the west side of 20th, where they maintained the quartermaster depot and mule remount station, or corral, which was built during Civil War times. We referred to this place simply as the "Corral," while Pierce Street, where we were located at first, was known later as "Bootlegger's Boulevard," apparently due to certain activities during the prohibition era.

In later years (1950s), I knew Carter Sheet Metal Co. and called on them in the vicinity of 13th and Grace streets, between Gate City

Steel Works and Eaton Metal Works. I believe this was the same family business. They were users of light gauge galvanized sheets; consequently, I did not sell them any steel.

So we have the story of a company from horse and buggy days, when steam was king, and the steel industry was definitely a growth industry, and concentrated in the Pittsburgh and New York areas. Andrew Carnegie was the world's richest man, and the railroad was the beginning, the middle, and the end of transportation—both for material goods and for people. Just imagine, this was not too long after the railroad across the entire country had been completed, and what once was a trip requiring several weeks was now down to a few days.

Drake's Brief History

It is likely that George Drake joined Mr. Wilson in the firm in 1889 or 1890, and Hugh Williams in 1893. John Morris Wilson was senior to George Drake by perhaps twenty years, and it is reasonable to assume that he was easing out and attempting to insert his son, Mr. J.M. Wilson, in his stead. Then, possibly, the junior Wilson decided he wanted to do something else with his career and left the company's employ. There is just no record or trace of either of the Wilson men after these earliest years, so that remains an unsolved mystery.

Drake evidently had some money to invest, and his name appears as a proprietor in 1893. The Williams name is not in the masthead until 1897. It is likely (but no record exists) that Hugh began buying capital stock from Wilson about this time.

We know that Drake came to Omaha from Springfield, Illinois, about 1889. We have his obituary from a Springfield periodical:

First shop building and office at 19th & Pierce streets. At left is John M. Wilson, founder. Horse and buggy driver is Hugh Williams. He called his horse "George Washington."

DRAKE—died at Omaha, Neb. Saturday, April 6, 1901; George Henry Drake, aged 40 years.

The death of Mr. Drake was sudden and no particulars have yet been learned here. The said intelligence came to this city through a telegram to his father, William Drake, the well-known boilermaker. Mr. Drake left on an afternoon train for Omaha, and the arrangements for the funeral will be made later.

George Drake was born and raised in Springfield, and is well known here. He learned the trade of boiler making in his father's shop and some twelve years ago went to Omaha, where he built up a successful business of his own, as a member of the firm of Drake, Wilson and Williams, and had become known as one of the substantial businessmen of the city. The crushing bereavement has come so suddenly that the members of the family have not yet had time to consider the arrangements for the funeral. At the time of his last visit to his old home during the Christmas holidays, Mr. Drake was the picture of health and his sudden death is a severe shock to his many friends here.

-*Springfield Journal.*

As to the cause of his death at only forty, Arthur Williams told me that Drake, first of all, never was really very active in the actual operations of the company, but was involved more as an investor, and also that he was a prodigious drinker, and was, in fact, a drunk, and that the circumstances of his death were that he was found in a downtown hotel, flat on his back in bed, having suffocated from his own vomit after a drinking binge. Whether this last part of the account is true or not, we will never know.

There is simply no other information to be had or to be found, either to corroborate or to debunk this account. No other record exists of which we are aware, and apparently no coroner's report was made.

Hugh Williams

Hugh was born in 1862 in Milwaukee and moved with his family to Chicago. Probably about 1875 or 1876. He attended public schools there and a business school in Chicago. He worked in the Union Stockyards of that city, becoming a hog buyer, from about 1879 to 1886.

About this time, he came to Omaha, and we have no record or any idea what motivated his move. Again, my cousin Georgia advances a story that she says she was told, and that involves a large shipment of furniture, and that he was paid to accompany the shipment to Omaha for a fee. This seems a fanciful account, and not really in character with old Hugh, but we will never know.

In Omaha, he went to work for Paxton and Vierling Iron Works, where he remained for six and one-half years, or until about 1892 or 1893. A later account of the circumstances of his move to Omaha from Walter Williams claims that he was accompanied by A.J. Vierling, which would explain his employment initially at that firm. Hugh was the first child born to William and Alice, of six total, all surviving birth and healthy, which was unusual for that time. However, Mary did die in her infancy, but we do not have the cause of death. The others were:

Mary, born January 9, 1864

William, born February 26, 1866

Charlie, born June 25, 1868

Walter, born April 5, 1870

Robert, born March 12, 1872

Here is the biography that appeared in *Omaha—The Gate City and Douglas County, Volume II* (S.J. Clarke Publishing Co., 1917):

HUGH W. WILLIAMS

Hugh W. Williams, president of the Drake-Williams-Mount Company, has been engaged in this business since 1897, and his initial step in this direction followed previous experience along the same line as an employee. He was born in Milwaukee, Wisconsin, June 30, 1862, a son of William and Alice (Roberts) Williams, who were natives of Wales. Each came to America in the early ’50s and settled in Milwaukee. Following their marriage, they removed to Chicago, where the father engaged in blacksmithing and horse shoeing, remaining in that city until called to his final rest in 1915, when he was in the eightieth year of his age. His wife passed away in Chicago, in 1908, at the age of seventy-three years.

Hugh W. Williams (the first), photo taken in 1926.

Hugh W. Williams was the eldest of their five sons. He pursued an education in the public schools and a business college of Chicago, after which he acquainted himself with the business of hog buying in the Union Stockyards of that city, being there employed from 1879 until 1886. He afterward came to Omaha and entered the

Paxton-Vierling Iron Works, with which he was connected for six and one-half years. He afterward spent ten years in the Wilson and Drake Boiler Works, later Drake Wilson and Williams, and on the expiration of that period, he organized the Drake-Williams-Mount Company in June, 1902. The business was started in a small way, but has grown to large proportions and in 1902 it was necessary to remove from the original location at 19th and Pierce Streets in order to secure larger quarters. Something of the development of the trade is indicated in the fact that there are now seventy-six employees outside of the office force and the plant covers two acres of ground. They are thoroughly equipped to take care of any kind of business having to do with tanks, smoke stacks and boilers, and excellent shipping facilities have been secured through trackage in their own yards.

In January, 1891, in Omaha, Mr. Williams was married to Elizabeth Shaw, a daughter of James and Katherine Shaw of Denver, Colorado. There are four children: Walter Hugh, who was born in June, 1892, was graduated from the Omaha schools and since that time has been associated with his father in business: Alice, born October 1, 1893; Georgia, born February 17, 1895; and Arthur, May 12, 1904. The three oldest have all completed the high school course.

In politics Mr. Williams is an earnest republican but not an office seeker. Fraternally he is connected with the Benevolent Protective Order of Elks. He holds membership in the Commercial Club and he is interested in all

> those forces and movements that have to do with the welfare and development of the city. There never has been an esoteric phase in his life record. On the contrary, everything has been open and above board and his honorable effort in business has been attended by most substantial and gratifying success. He ranks high as a leading manufacturer and citizen and is well known as a representative man of Omaha.

As stated in the bio, Hugh's father and mother, William and Alice (Roberts) were both natives of Wales. William was a blacksmith by trade, and both came to America in the mid 1850s, at young ages. William was born in 1836, in Llanfaelog, County Anglessy, in Wales, and Alice in 1835 in the same place.

At this point, I would like to express a huge amount of gratitude on behalf of all Williamses for this decision and suggest that all of us who wound up Americans can thank this man and this woman for having the courage and the determination to set out for a brand new home! Indeed the Williamses owe William and Alice a huge debt.

William passed away in Chicago in 1915, but Alice preceded him in death in 1908, at age 73. We recently learned that he is not buried next to Alice in the cemetery, which could mean that they were divorced, or perhaps separated. He had a blacksmith business all of those years in Chicago.

My wife, Mary Jean, and I visited Wales a few years ago and found Llanfaelog in Anglessey, where William and Alice were born. There is a small post office and a community services building, plus a small Methodist church and cemetery along the highway, and that is all there is to the town. I found the cemetery much overgrown with weeds and grasses, but I could see that the names Williams and Jones were seemingly on every stone marker, and some with both names on the same headstone, Williams-Jones. A few Evans and Davies, but Williams and Jones were far and away the most common.

Another small diversion at this point concerns a told and retold and then told again story (or myth?) that has been around for a long, long time. It concerns the very name, Hugh Williams. I first read a similar account of this story in the *World-Herald*, in a feature they used to carry, "Believe It or Not," by Robert Ripley. Here is a version that was taken from the *World-Herald* in February of 1890.

```
The Name a Mascot
Men who Bore it and Escaped All Sorts
of Perils.

Leeds Mercury: In the year of 1664,

On the 5th day of December, the English ship
Menai was crossing the straits, and capsized
in a gale. Of the eighty-one passengers on
board but one was saved: his name was Hugh
Williams. On the same day, in the year 1785,
a pleasure schooner was wrecked on the Isle
of Man. There were sixty persons on the boat,
among them one Hugh Williams and his family.
Of the three score, none but old Hugh Williams
survived the shock. On the 5th of August,
1820, a picnicking party on the Thames was run
down by a coal barge. There were twenty-five
of the picnickers, mostly children under 12
years of age. Little Hugh Williams, a visitor
from Liverpool, only 5 years old, was the only
one that returned to tell the tale. Now comes
the most singular part of this singular story:
On the 19th of August, in the year of our
Lord 1889, a Leeds coal barge, with nine men,
foundered; two of them, both Hugh Williams,
an uncle and a nephew, were rescued by some
fishermen, and were the only men of the crew
```

who lived to tell of the calamity. These are facts which can be substantiated.

Over the years, I have seen several different versions of this yarn. There is always quite a little variation in the details. People have run across these and have sent them to me. I took along one of these stories, in print, when Mary Jean and I made the crossing of the Irish Sea from near Dublin to Anglessey, in Wales. I had fun showing it to several people on board and warning them of the great "hazards" of the trip. I don't think they became too fearful.

Wilson Steam Boiler Works

Some "reading between the lines" becomes necessary at this point, because we do not know how Drake and Hugh Williams got together, or what motivated them to toss in with John Wilson in his boiler and heavy sheet metal business.

No doubt Hugh had learned quite a bit about the steel business, some of course from his father, the smithy, and some from his association with Paxton and Vierling Iron Works. We know that George Drake was a young man who loved his liquor. We also know from family tales that Hugh was fond of a pint or two and was quite a ladies man to boot.

Sometime around 1892, perhaps, there was an encounter in one of the local watering holes. They had things in common. They were in related businesses, they were "of an age," and they both had a taste for the local brews.

It seems more than a little feasible that they would meet in such a place and that they would form a friendship based on their mutual interests, even though Hugh was newly married in January of 1891.

Hugh had a record all his life of carousing around, in spite, no doubt, of Lizzie's consternation.

What brought them to Omaha? We just don't really know in Hugh's case. He had training in the hog buying trade, but never seemed to take that up in Omaha. In George Drake's case, could it be that he had sort of worn out his father's welcome in Springfield, and maybe his father decided that he should take his act (including his drinking and carousing) on the road and go and make his own way in the world? Very feasible, and maybe he had a tidy grub stake with him, and just maybe Hugh could see this as the means to get into business for himself and to begin to be his own boss. We know, certainly, that he had the smarts and the drive to go it alone, and he had the spirit of the entrepreneur.

Cousin Georgia Lee states that Hugh had a blacksmith shop in Fremont, Nebraska, along about this time. In fact, she claims that her mother, Alice, told her that he had three such shops, although where the other two were located, or any of the shops situated, there is no record. Apparently, he gave these up because of the difficulty caused by the proximity to Omaha. We know there were no paved roads at this time, so that must have been quite a trip. At any rate, these sidelines, if they were real in fact, were not maintained once he made the move to join the Wilson firm, but it is possible that they got him enough money to buy an interest in that company. There is absolutely no further evidence of his having these blacksmith shops, and it casts great doubt on my cousin's story.

Conjecture

Because of his association with Paxton, Hugh probably knew of John Wilson and his business on Pierce Street, and, as the old story goes, "with your money and my brains..." no doubt a plan was hatched, and a deal consummated. They knew that John Wilson was not getting any younger, or that perhaps he wanted to retire to something else, so they probably approached him, but no record exists of an agreement of any kind, or a price, shares of stock—nothing. Of course at that time, it was not organized as a corporation.

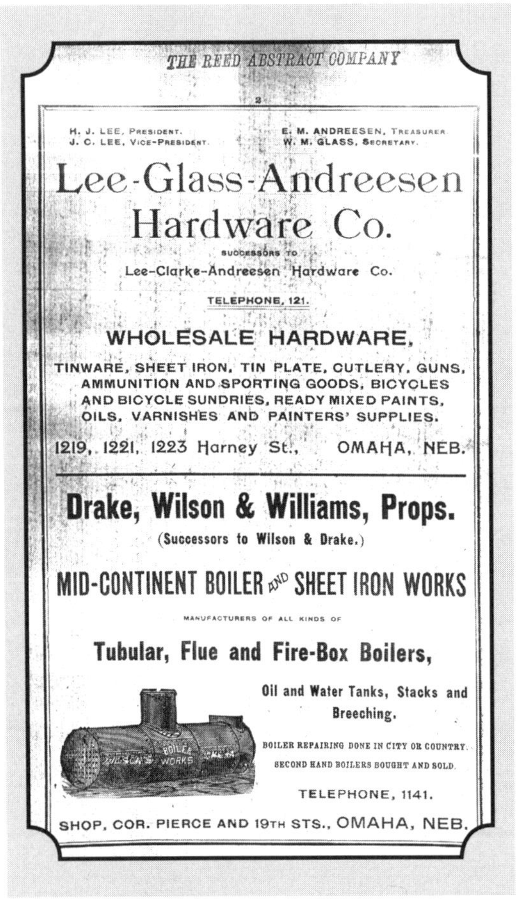

Ad, about 1899, from McAvoy City Directory.

Drake, Wilson and Williams

So they make a deal, probably in 1893, and by 1897, we see the Williams name for the first time in the ad in the McAvoy City Directory. Hugh is eclipsing George, and John Wilson, for that matter. Still, there does not seem to be much actual growth in the company in those years. I'm sure Hugh was learning about boilers and about the tank business from John.

The company did build boilers from the ground up, and a few were still operating in several Omaha locations up until a few years ago. Apparently, as builders, they were not large enough or equipped to really compete with the large Eastern manufacturers of boilers

Shop crew at 19th and Pierce shop, circa 1900.

such as Kewanee, Cleaver-Brooks, Pacific, and others. The company was probably very much undercapitalized to mount a fight for the market. We have no idea of how, why, or when the Messrs. Wilson bowed out of the business, but of course, we know about the untimely death of George at about forty years of age.

Another scenario is possible, however, which would have Hugh not succeeding to any ownership until 1897 or early in 1898, which is when his name appears in the firm's title (Drake, Wilson and Williams). He probably made a deal with John Wilson to buy him out, because this (1897) is the last time that Wilson appears in the firm's official name.

Surely, with Wilson bowing out, and with George no doubt being little help, it is apparent to Hugh that he needed some additional management expertise. Soon, all hell would break loose. Omaha is going slightly mad now. Remember that the city is really busy and excitement is building to a frenzied climax with the huge Trans-Mississippi Exposition going on in 1898. Forty states and ten foreign countries took part. The roof is blowing off—Omaha is definitely a boomtown. A dusty frontier town is threatening to become a metropolitan city.

According to Frederick Luebke, retired University of Nebraska–Lincoln historian, "The 1880s were a decade of enormous importance, because it is by 1890 that Omaha starts to take shape." Omaha was developing heavy industrial jobs with railroads, smelters, stockyards, and packing houses, which appealed to the waves of job-seeking immigrants, many from Southern and Eastern Europe. Development of the packing industry gave Omaha a working-class culture that persisted deep into the twentieth century. "It all jelled in the 1880s," Luebke said, "and Omaha was a viable rival—at least in population—to Kansas City."

The super-heated economy and building boom consumed 150 million bricks a year, produced by fully fifty-two brickyards. One must remember that bricks were used for most of the streets that were being laid down.

Hugh Gets a New Partner

What must it have seemed like to Hughie when his partner takes the gas, and they find him toes up in that downtown flophouse? Sometime during this next year, he finds his answer in the person of Mr. Ernst O. Mount—just the kind of stable, steady, and no nonsense guy he needed. Furthermore, he knew Mr. Mount from some years earlier, when he worked for the Wilson concern before he went to Denver. Mount was of good German stock, born on a farm in Illinois. His father and mother (Grobe) were born in Germany. He was about thirty years old and apparently was able to buy at least some degree of ownership with Hugh when he came to work. They made a good team, and now Hugh was ready to step out and do some growing.

Mount was no stranger to the boiler and sheet iron business. Here is his brief bio, taken from his obituary, done in his own words, which he wrote on the way to the hospital in 1935, and which was to be his last trip, and where he died at the age of 63:

> He walked from Jacksonville (Ill) to Council Bluffs in 1884 and went to the Wilson Steam Boiler company at the age of 12 for $4 per week, on March 22, 1884, and stayed until 1896 when he went to Denver to the Shaw Boiler company and Denver Boiler and Sheet Iron Works for five years, leaving there in 1901.
>
> He came back to Omaha and went into business with Hugh W. Williams, then employing eight to ten men. The business grew until the present day the concern is employing from 50 to 60 men, and

Ernst O. Mount, president, in 1931, until his death in 1934. He wrote his own obituary on the way to the hospital during his last illness.

> have had as many as 150 during the war. He was never out of work, and was a self-made man, having had only one year of day school. The rest of his education was attained after working hours at night. In those days, they worked 59 hours a week.

It was certainly likely that Hugh, knowing Mount from those earlier times, sent for him in Denver with an offer to have some ownership in the newly incorporated firm, and that he was glad for the opportunity to work once more with Hugh. It also seems apparent that he was a very key figure in this early growth of the company and a stabilizing force.

We have a limited glimpse into the tenor of the times with a look at one of the oldest payroll records that we have been able to find, which was for the week ending August 16, 1902. There is a list of seventeen names, headed up by H.W. Williams and E.O. Mount drawing the princely salaries of $18.00 each. The men worked a regular ten-hour day, and the normal week was six days. Wages ranged from 15 to 30 cents per hour, except for the last entry which was "new kid," and his rate was 12 and one-half cents per hour. The total payroll for the week was $191.50.

These had to be very exciting times, and to be a relatively young man beginning an all but new venture at 23rd Street and the Union Pacific Railroad had to be daunting to Hugh. He now has a wife and a ten-year-old son, Walter, two daughters, Alice, age nine, and Georgia, seven. There are mouths to feed, responsibilities, and plenty of pressures to make a go of the business. His six and one-half years with Paxton and Vierling Iron Works and ten years with J.M. Wilson had qualified him and lent him the valuable experience that he needed.

The years that followed were good ones for the company, and we grew stronger. We needed to, to weather the great money panic and recession of 1907. Basically, it was a panic that was the result of rumors that spread about the solvency of key banks. A run was started that

23rd and Hickory yard, looking north from UPRR main line. Stiff-leg cranes were the only means of loading and unloading rail cars, early 1900s.

spread quickly across the country. Teddy Roosevelt tried to assure the country that the system was secure, but the crisis caused the Congress to convene hearings, the culmination of which was the actual creation of the Federal Reserve System in 1913.

One account says that they made the move from Pierce Street to 23rd and Hickory to avail themselves of more room, but it was about this time that the B & M Railroad (Burlington and Missouri) built a viaduct that spanned Pierce at just about the site of the old plant. They were forced to move because of condemnation, but it is clear that they did need the added room to grow. Art told me that the main structure, which was wood frame, was moved all that way on wooden rollers and pulled by several teams of horses. This would have been the old East shop building, which had a dirt floor, and which was never set "square with the world," and which always had more broken out windows than

Early 1900s interior of old wooden shop, 23rd and Hickory.

intact ones. Really, about all the building afforded us was a roof over our heads—and not a very sound one at that.

Arthur Williams "On the Record"

In about 1970, I had the foresight to get my father, Art, to sit down for a fairly extensive interview on tape, although not as long as I would have liked. Art began by saying that he started work at the plant when he quit school in about 1924, and his pay was $2.50 per week to start.

He remembers spending a lot of time rethreading bolts and sorting rivets. These were bolts and rivets that had been reclaimed somehow from jobs, and though this operation would hardly be profitable today, back in the twenties it was a necessary, if rather tedious, job.

He spoke of old "Dutch Charlie" (real name Luessner) who was the teamster for Drake-Williams, and who lived right across the street from the Hickory Street plant, and who had a stable in the rear. He had two horses and a wagon, and he hauled material that we fabricated for assembly in the field. At this same stable, he also kept Hugh's horse, "George Washington," and a carriage, for the daytime, until Hugh was ready to use them for the drive home. He mentioned that we had a lot of work out at the Florence Water Works about this time, and we were building coal bunkers, oil storage tanks, and filter tanks—all of which had to be riveted together. In those days, we were also making complete boilers, and again, these were fabricated in our shop but totally assembled in the field. Materials were loaded on the wagon using three guyed derricks, or stiff-leg cranes.

Arthur Williams, about 1925.

Earlier, we employed around 110 men in the Hickory shops, and an additional 70 or so in the shop we owned at 20th and Center streets.

These were the war years, around 1916. The first wood shop was the one we moved (dragged) up from 19th and Pierce, and then we built the "brick" shop about 1912. After the wood shop was re-sited, we added about 100 feet to the south, making a total shop of about 225 feet. There is some question about which railroad caused us to move—the Great Western, according to Art, or the Burlington and Missouri, as per other records and maps.

After this move, the company had to buy two additional narrow parcels of land to accommodate the brick shop and yard area between.

To the question about the earliest electric arc welding at our plant, Art said this came about in the early thirties, when we undertook a job for Nebraska Power Co. (later the Omaha Public Power District). We furnished and erected a very tall stack as well as breechings and associated equipment, and electric welding was specified. After this, welding caught on rather quickly, although there was still riveting on much of the work. This welding was the bare wire process, and a lime powder flux had to be employed by hand, using a funnel-like device on top of the arc. He recalled that our first welder was a guy by the name of Virgil Birch—a man that we got from Eaton Metal Products Co.

Art talked about the extensive work during the World War (it was not called WWI, after all, since there had not yet been two wars) for the potash industry, and for the sugar beet companies, Great Western Sugar and American Beet Sugar Co. We made evaporators, condensers, and dryers at this time. The potash dryers were large: 6 feet in diameter and possibly 60 feet in length. The material had to be pulverized as well as dried, and for this operation, lengths of chain were welded to the inside so they would thrash around and break up the material when the drum was rotated. We also did much of the necessary machining, together with the old Melchior Machine Works in Omaha.

We discussed Teddy Evert (his story appears later in this book), and Art remembered him well. When we built the large stack for American

Smelting & Refining, which for a time was the tallest self-supporting stack in the world, Ted Evert was the man in charge of the field erection.

For many years, in the main office on the east wall was a framed photograph of an old gentleman, known only to a very few of our people. This was, Art reminded me, Alex Scully, and he was a close friend of Hugh's from the old days and was one of the founders of United States Steel Supply Co., the warehouse arm of U.S. Steel itself. Art confided that he was of very great help to Hugh credit-wise, helping the young company during some rough years and some tough sledding.

During the war years, steel was very difficult to come by, just as happened in World War II. We established valuable mill relations with the young Inland Steel, before that mill was quite so large.

This was about all that was covered in this interview, and, of course, I wish that we had covered a lot more about personalities, those of Hugh, and those of Ernst Mount, his brother-in-law Frank Wagner, and Ted Evert. But I am grateful for the tape recording, which is better than nothing at all.

Omaha's Coming Out Party

That these were "high times" in the city may be an understatement. Consider: In 1895, the Knights of AkSarBen, the nonprofit organization, is founded by Omaha businessmen. (I have in my possession a certificate for five shares of stock in the original AkSarBen Exposition Company, issued to Hugh Williams.) This was a civic endeavor by leading Omaha businessmen to promote the city and environs and to encourage the development of commerce. They wanted to promote a Mardi-Gras–like annual event similar to New Orleans. Eventually, when the Knights of

AkSarBen sold the property to the county, there was a big question in my mind as to why they, and not the community of Omaha, really owned that property.

More high times when, in 1898, at noon on June 1, President William McKinley declared the Trans-Mississippi and International Exposition and Indian Congress open via electrical current transmitted all the way from the White House.

However, Omaha was not alone in its bid for the exposition. Minneapolis, Chicago, Kansas City, and St. Louis were all interested when the talks began in 1894. The fame of William Jennings Bryan, the Great Commoner, was at its height when the Trans-Mississippi was hatched. When the Trans-Mississippi Congress was convened in St. Louis, Bryan convinced the group to meet the following year in Omaha. Bryan and others were prepared, and the legwork was done.

Edward Rosewater had founded his newspaper and had become an influential political leader of the West. Gurdon Wattles, early an influential leader in Omaha, was also instrumental in getting the event staged here. He was the financial wizard of the expo, the tactics of which were later called into question by many. The "scheme" was the sale of stock to finance much of the effort, but investors did realize about 90 cents on the dollar, so it was not really an out-and-out scam.

The Exposition was held in north Omaha on a tract of land owned by the Kountze family, and on the site of the old fairgrounds between Laird and Boyd streets, from 16th to 20th streets. McKinley paid a visit to Omaha and the Exposition on October 12, the day of the armistice that ended the Spanish-American War. It is a bit of irony that a little later, McKinley was assassinated by an unemployed steel worker and gave way to Theodore Roosevelt. The Trans-Mississippi was, and still remains, the largest public event ever staged in Omaha. This was in spite of the biggest business recession in twenty years, and despite drought and crop failures in the area. It was said that as many as 5,000 houses stood empty, but the economy was on its way to recovery by 1898.

Not nearly as well known was the Greater American Exposition held in 1899 during the summer, on the grounds of the Trans-Mississippi Exposition. It focused on U.S. colonialism, the result of the Spanish-American War, and featured exhibits from the Philippines, Cuba, and Hawaii.

Another notable event, and one that attracted national attention, was the kidnapping of Edward A. Cudahy, Jr., the infant son of the major Omaha meatpacker. Pat Crowe, a former employee at the Cudahy plant, was the kidnapper and was paid a ransom of $25,000, and the boy was returned. The Omaha jury that acquitted Crowe seemed to not consider the question of guilt, but seemed to say, "Isn't it okay to rob a member of the beef trust if you can?"

Steam was king. It powered the ships that plied the waters of the Missouri, it ran the huge locomotives on the newly laid tracks around town, and it likewise supplied the power and heat for all the processing in the great packing plants sprouting up in South Omaha, or "Magic City" as it was called. It was estimated that in the early 1900s, 80 percent of Omahans had livestock-related jobs. The boiler and tank business was definitely the place to be—it was the "happening" place.

However, something that possibly did not receive a lot of attention along about this time took place near Beaumont, Texas. An oil well was brought in—a huge gusher that spewed oil 200-feet into the air and blew for nine days before workers could cap it. This turned into a bust, until years later, when the black gold was recognized as being valuable for powering ships and locomotives. No one could hear the approaching footsteps, or feel the challenge to steam power, that was mounting in the not-too-distant future. Apparently the owners and operators of Drake, Wilson and Williams did not worry too much about these new forms of power and energy—steam was good enough for them!

Early Omaha Industry

One of our oldest and best customers, American Smelting & Refining Co. (ASARCO) is mentioned in *History of the State of Nebraska*. This was a publication of the Western Historical Co., A.T. Andreas, Proprietor, 1882. It was reproduced in 1975 by Unigraphic, Inc., 1401 North Fares Ave., Indianapolis, Indiana, 47711.

ASARCO is listed simply as "Smelting Works," this being in 1882. It is referred to as "by far the largest refining and smelting works in America." Officials of the company are shown as W. Mead, Omaha, President; H.W. Mead, Quincy, Ill., V. President; E.W. Nash, Omaha, Secy-Treasurer; Charles Balbach of Omaha, Suptd. This last man may have been the founder of the Balbach Company, suppliers of oxygen and industrial gases, which later became Air Products and Chemicals Co. in the late 1950s.

The facility became American Smelting & Refining Co. in 1899. The company has sustained a lot of abuse in the press and in city hall for air and ground pollution. In my years of association with ASARCO people, I knew them to be very professional, principled, and fair in their dealings with the public—suppliers and customers alike. They were ultra safety conscious many years before OSHA came along. They have been good citizens in every way, supporting local civic projects, charities, and events. I believe they thought that they were making every effort to be environmentally safe, and that they have been judged unfairly, with standards that have changed radically from the standards in effect during those past years. It is a totally different set of ideas, and our newer science has been unfairly applied in a retrospective way to their operations.

Another early company listed is Omaha White Lead Works, formed in 1878 by noted early industrialists, W.A. Paxton, C.W. Mead, Levi Carter (Levi Carter Lake was named for him) and S.E. Locke. They availed themselves of "pig lead" from the "Smelting Works." White lead was used for making waterproof seals in cast iron pipe, leading in glass in steel windows, and joining together sheets of tin (ceilings) and gutters and downspouts.

Pacific Storage and Warehouse Co. was an entry, and included was a description and picture of Charles H. Gratton. This is the grandfather of Dick and Geody (George) Gratton, except that their father, George, was adopted by Charles, so therefore they would not be actual blood relatives.

Another Omaha company that was growing up with ours was Paxton-Mitchell Co., formed as a partnership by Thomas Mitchell and James L. Paxton, Sr., in 1898. Their outstanding invention was developed by Mitchell—metallic packing material for steam engines. The entrepreneurs incorporated in 1901, about the same time as Drake-Williams-Mount Co., with $60,000 and one employee at 13th and Howard streets. The company moved to its present location on Martha Street in 1912. James L. Paxton, Jr., took over from his father in 1937.

During World War II, Paxton-Mitchell produced metallic packing for 40 percent of the country's steam locomotives and metallic castings for 60 percent of the famous "Liberty Ships." Paxton-Mitchell faced similar challenges to what we did, with the eclipse of steam power by other forms, such as diesel and electric. Now, Paxton-Mitchell makes and sells hydraulic castings worldwide and produces specialty products for railroads.

The company also has a machine shop operation at 21st and Pacific streets and a manufacturing unit in Blair, Nebraska, making "snooper" cranes for use on bridges for maintenance work. This facility is Jebco Division that was formed in the mid 1950s, and that purchased the "Elegrader," which was developed by us and engineer Verne Reisser. More about the Elegrader follows later.

The Horseless Carriage

The automobile had not really arrived as yet on the scene. Hughie, according to my father, had a faithful horse by the grand name of "George Washington" that squired him around town in his fancy buggy. Art said that the horse knew his nightly route from the plant to his home by heart, and he stopped automatically at several watering holes on the way for Hugh's liquid refreshments. It was the year 1900 and 1901 that Ransom Olds developed the first Oldsmobile in Lansing, Michigan, and it became the first commercially successful car in this country. This inspired the hit song, "In My Merry Oldsmobile," and 600 of them were sold in 1901.

What about the automobile and Omaha? The following lines, originally from the *Omaha World-Herald* and related to me from columnist David Harding, are these:

> The horseless carriage first appeared on Omaha streets when merchant Emil Brandeis paid the ungodly sum of $5,000 for an automobile delivered in 1895.
>
> Otto Baysdorfer didn't have that kind of cash, so he and his two younger brothers did what

any self-respecting mechanical geniuses would do—they spent two years designing and building their own. In 1898, they rolled out the first car ever built in this area, the "Ottomobile."

It was an incredible feat because the Baysdorfers had no off-the-shelf parts to work with—no transmission or piston or spark plugs. All were fabricated locally from their drawings. Otto, Gus and Charles were the sons of a cigar maker who brought his family here from Davenport, Iowa, around 1887. The brothers ran a bicycle hospital, as it was called, and fixed electrical motors for a living. In their spare time, they introduced this community to an array of the 20th century's early marvels.

Harding went on to talk about how they fitted a small engine to a bicycle, and literally built a motorcycle, and further, how they built an "airborne wonder," the dirigible. Their wives and friends sewed the gas bag from hundreds of yards of Japanese silk, which was dipped in linseed oil and coated with varnish as a sealant. A frame under it ran the entire length of the airship, an engine propelled it, and the pilot controlled its ascent or descent by moving forward or backward along the frame.

They named it the Comet, and it entertained Omaha crowds several times in 1907. Charles, the family dare-devil, took the craft aloft at Ak-Sar-Ben's fall festival, when a strong wind carried him away to the south along the Missouri River. He moved forward and brought it down just above the treetops.

> When he saw an old man fishing along the riverbank, he dropped a rope, and asked the man to grab it and slow the craft down. The man told him to catch his own rope, and then went into his cabin. Further down the bank, several fishermen caught the anchor line, and the dirigible landed safely.
>
> The Baysdorfers were far from finished with tinkering and inventing. Later, they met with Glenn Curtiss when he came to Omaha, and they learned everything that they could about airplane construction. They then built an airplane and flew it a hundred or so feet from a field near Waterloo.
>
> They introduced the city to all kinds of devices—from wireless radio to portable x-ray machines—and they invented a few of their own. The Affinity spark plug was probably their most profitable creation. Others included a gasoline gauge, a gas lamp and a plug-in electrical socket.
>
> They could not refrain from all sorts of bizarre inventions, such as a device for buttering toast and a self-setting mousetrap!

It was later, in 1908, that Henry Ford, vowing to build a car for the multitude, launched the Model T. Within one year, 10,000 of them were putt-putting across America, and it was a real manifestation for all that the machine age had arrived. The world heard about the first production line, and we knew the term *mass production*. It was then that Henry declared that you could choose any color car that you wanted, as long as it was black!

Telephone service also had been established in Nebraska and Iowa. It was said of Hugh that he really didn't need one, but that he was so used to "hollering" that he could walk the short distance to the door of the office and shout all the way downtown, to Johnson Hardware, for example, if he needed to order some bolts, or whatever. They could hear him perfectly. Of course, Johnson Hardware is also one of Omaha's oldest established companies.

Perhaps this is why Art, much later, and not getting along famously with his brother, coined the name for Walt that went with his (W.H.) initials, "Whoop and Holler!" We have no record of what Walt called Art behind his back.

Other industry prospered in Omaha. The ranchers and farmers who brought their cattle to sell in South Omaha needed diversion. The great meat packing houses came, and South Omaha became known (almost officially) as "Magic City," as Swift, Armour, Wilson, and Cudahy built huge packing plants there. The others were satellite plants from the Chicago area, but Cudahy, as we have seen, was headquartered in Omaha.

But first, in 1883, a cattleman of Scottish descent named Alex Swan bought land in the South Omaha area and suggested the place be called "New Edinburgh." Swan had large herds of cattle in Wyoming and joined with Omahans William Paxton and John Creighton in buying land and constructing the stockyards that would become the largest in the nation. This was the Union Stockyards Co., which was begun in 1884.

This activity ignited a population boom, and land values likewise rose. It was a rugged beginning, and by 1886, the town had no churches, no jail, and only a ramshackle schoolhouse. So, it seems odd that Hugh Williams, arriving in Omaha at about this time, and having had experience in the stockyards of Chicago as a hog buyer, did not wind up in this business. It remained wild and wooly for many years, and it was not until 1915 that it was merged into Omaha. This event was not popular among the citizens of South Omaha, but was inevitable due to the vastly larger population of Omaha.

Omaha – Wide Open Town

The poet, Carl Sandburg, long ago wrote:

"Omaha, the roughneck, feeds armies
Eats and swears from a dirty face.
Omaha works to get the world a breakfast."

Diversion for the farmers and ranchers was in the form of the ornate two-story Diamond Gambling House operated at 1312 Douglas Street, until city officials closed it down in 1893. And down the street at 912 Douglas was a famous house of prostitution run by one of the era's widely known characters, Anna Wilson (no relation to John M., one hopes). She willed the house to the city following her death in 1911. Its stone porch columns of nude women were taken down and replaced with standard wood posts. We have no record nor rumor to indicate whether or not Hughie ever availed himself of these attractions. By 1910, it was estimated that Omaha had 2,500 prostitutes working at fully 100 brothels. There were pockets of poverty, however. Squatters built shacks north of the downtown area, using materials salvaged from the city dump. These neighborhoods had names, such as Vinegar Flats and Blind Pig Alley.

Omaha's emphasis, conscious or not, was on building heavy industry, with jobs in the railroads, the smelters, the packing houses, and the stockyards. Until the 1880s, Omahans traveled on dirt streets and drank well water or Missouri River water that had been allowed to settle.

While Omaha was booming, it was not quite as spectacular as the census of 1890 tried to show. That census purported to describe a growth from the 1880 figure of 30,518 to a whopping 140,452 in

1890. The census was a colossal fraud, but the state's congressional districts were increased from three to six, which was a decade earlier than Nebraska deserved.

So, we see that Omaha was wide open in those formative years and between 1900 and 1930 experienced a doubling in population, the greatest growth before or since that time. Small wonder that author David Bristow was inspired to write his historical accounts of our town in his book, *A Dirty, Wicked Town*.

Business was good for the company, too, even before the move to Hickory Street. One of those brickyards previously mentioned was owned by John Kiewit and is believed to have been right across Pierce Street from our plant. This Kiewit would be the brother of Peter Kiewit, the Peter Kiewit who was the father of the Pete Kiewit that built that company into such a giant among contracting firms.

What did Hugh make of life in the beginnings of the twentieth century? He had seen rail transport speed up travel by a factor of at least ten over the horse and mule, had seen the advent of telephone and telegraph make messaging almost instant, and balloon cars become the subject of almost daily experiment, meaning that travel over any kind of terrain would become possible. Another increase in travel speed at that rate (factor of 10) would mean that man might well be traveling 600 miles per hour! All of these accomplishments and improvements must have helped to encourage Hugh and his associates as they looked to the future. No wonder they stepped out and took the risks to get ahead.

As far as George Drake is concerned, there is no evidence and no reason therefore to believe that he was related to Robert Drake of this era, who owned the Standard Bridge Co. and was a noted builder of bridges in the area. But he was the father of William Drake, developer and builder of the Drake Court Apartments in Omaha. This Drake was once owner of some 1,100 apartments and wanted to become the largest apartment owner in the world. It is believed that the money necessary to finance the start-up of the Drake Court Apartments by William was obtained from the proceeds of the sale of George Drake's stock in our company, after his death.

However, economic conditions inevitably turned against him, and he found himself competing against his own apartment houses. In 1925 he was forced to declare bankruptcy.

Trans-Mississippi Exposition

A few more words about the Trans-Mississippi Exposition before we move on: The chief architect was the late Omaha architect, Thomas Rogers Kimball, who also designed some of the finest buildings in town. Kimball was born in 1862, the same year as Hugh, in Linwood, Ohio, and moved to Omaha with his parents in his early teens. After graduating from high school, he attended the University of Nebraska for two years but did not graduate. He then entered the Massachusetts Institute of Technology where he studied architecture until 1887 but did not graduate from there either; however, he was given an affiliation with the School of Architecture. Kimball then moved to Paris where he spent a year studying at the famous L'Ecole des Beaux Arts. He went to Boston in 1888, and married Annie McPhail the following year.

In 1891, he formed an architectural firm with MIT instructor C. Howard Walker and Herbert Best. Later, Kimball moved back to Omaha and opened an office with Walker. His appointment to be architect-in-chief for the vast Trans-Miss project was no small feat, considering that he was not yet thirty-five years of age.

His associations in the East, and in Paris, made him unusual. There were not that many architects who had studied in Europe, and he began attracting many high-profile projects, including St. Francis Cabrini Church and the Burlington Station. The latter structure was heavily altered in 1930, and it is interesting to note that its Doric columns were removed and taken to Lincoln where today they line the walkway near the east stadium at the University of Nebraska–Lincoln.

Among his many designs are the Omaha Public Library, the Fontenelle Hotel, and St. Cecilia's Cathedral. He did not live to see the completion of the latter, which was not finished until twenty-five years after his death.

It is said that he hunted buffalo as a boy, with Buffalo Bill and General George Crook. He attended Omaha High School, now Central High School, and must be among that school's most notable graduates.

His firm was unquestionably the primary firm in Omaha. As his circle of contacts grew, he attracted the attention of Gurdon Wattles, prominent Omaha banker, who selected him to design his home in 1894. Then it was a natural process of events for Kimball and Walker to be named co-architects in charge, after Wattles was named president of the Trans-Mississippi and International Exposition in 1896. They spent two full years working on the project. They were responsible for overall site development, including perimeter buildings. They designed several major buildings, some smaller structures, and the imposing Arch of the States, at the main entrance.

The buildings were constructed of strips of wood covered with a mixture of plaster and horsehair and were temporary by design, which meant the structures could be built at about half the cost of permanent buildings. The site consisted of some 184 acres and included numerous elaborate, although temporary, buildings.

A vast lagoon or grand canal ran the length of the property. The Grand Court, the centerpiece of the exposition, was bounded by 16th and 24th streets, and Pinkney and Pratt streets north and south. This project gave Kimball a lot of notoriety, and he became one of the big hitters in architecture. He went on to design the Electric Building at the 1904 World's Fair in St. Louis.

By 1918, he had gained tremendous stature among his peers and was elected president of the American Institute of Architects. In 1927, he went into partnership with architect William Steele. He maintained a successful professional life as well as a well-rounded home life, playing golf and teaching painting classes.

Others had at least temporary financial bonanzas, such as Minna and Ada Everleigh, who opened a brothel and supposedly doubled their investment over the five-month run of the exposition. They moved on to more fertile fields in Chicago after that. Pickpockets were rampant according to the *Trans-Miss News*, the official publication of the association. A myriad of side shows, amusing attractions, and novelties were among the 245 concessions.

Little Egypt, the "hottest hootchie-cootchie dancer outside the streets of Cairo" garnered a lot of attention. Another was Chiquita, the Living Doll. The Cuban midget was said to have twenty-one trunks of costumes and jewelry.

Omaha's combined hotel capacity exceeded 50,000 rooms by the time the expo opened. Another 100,000 beds were available in boarding houses and private rooms. Rents ranged from 50 cents to $5 a day, with most between $1 and $2.

After the exposition was over, succeeding generations of Omahans have wondered, "Was it all a dream? Was it reality?" Since none of the structures were permanent, and since many of the hopes and economic benefits did not materialize, the whole event seemed a mirage. This was to bring Omaha national praise and recognition, and Omaha was to "arrive" due to this effort.

The reality, however, was far less. Even though the fair was judged a financial success for its promoters, there was little long-term economic benefit to the community. When the Chicago House Wrecking

Company finished demolition of the grounds in 1900, it was as if a traveling trade show had folded, All that remained were the ribbons and banners—and the cleanup. One can imagine the feelings of Thomas Rogers Kimball and his associates at this time.

Drake-Williams-Mount Co.

Meanwhile, after Drake's sudden death in 1901, the plant and offices, as we have seen, made the move to the present location at 23rd and Hickory. Apparently, more people were needed in the new location, and were added. The company was officially incorporated for the first time, and with a new name, Drake-Williams-Mount Co., in 1902. The original articles of incorporation state: "The nature of the business shall be the manufacture, repairs and sale of all kinds of steam boiler, steel tank, sheet iron work, and everything pertaining to a general boiler works." This was the sole nature of the business for many years.

A picture of the "new" plant at 23rd and Hickory, identified as "summer 1904," taken in the yard outside of the shop building, shows nineteen workers, one of whom is apparently a foreman, or boss. Probably a few men were missing from the picture, so one would imagine that the shop force consisted of about twenty-four or so. The wooden building, of course, is the one that was hauled over from 19th and Pierce. Lots of steel shells are scattered around the yard, indicating that business in stacks and tanks was brisk.

This was the era of the standpipe and water tower for the company, as we will see with Ted Evert's story of his South Dakota job in 1907. We were doing quite well in spite of the unsettled business conditions and the banking crisis at the time.

23rd and Hickory yard showing rolled sections for tanks or stacks, early 1900s.

Another picture, labeled "Summer, 1902 19th and Pierce" shows fourteen men (none looking too happy, by the way), with one on the left probably Frank Wagner. Next to him is another man dressed in vest and tie and is not, contrary to the labeling on the back of the photo, Ernst Mount. At least it does not bear any resemblance to other pictures of Mr. Mount.

Ernst O. Mount not only joined Hugh, but came in as a principal with invested capital. Again, records are lacking, but it is supposed that he was sold the interests that had been George Drake's. Mount proved to be a very good addition, and a valuable man for the company. He was, as we have said, a no-nonsense type of guy, and a believer in a top-down style of management.

My father knew him well and maintained great respect for him. When he spoke of him (not nearly often enough), he always referred to

Shop crew at Hickory Street after the building moved in 1904. This grew to nearly 150 men in the following ten or twelve years.

him as "Mr. Mount," not "Ernst" or certainly not "Ernie." He told me once that Mount liked to sort of stroll around outside, to the rear of one of the shop doors, and then pop in the door kind of unexpectedly, to see if he could spot anyone loafing or doing something other than just working. Not exactly akin to participative management, but the old style, lay down the law, just nuts-and-bolts kind of management. Apparently, Mount was the one that Hugh depended upon for shop operations, while he carried on the administrative, the contracting, and the regular business affairs.

In 1902, the nation knew nothing of a fledgling company incorporating out on the "frontier"—Omaha, Nebraska—but the story of that year was the ending of the Boer War between Britain and South Africa, in which Winston Churchill first came to prominence as a young officer in the British Army.

The teddy bear (after Teddy Roosevelt), animal crackers from Nabisco, and the Tournament of Roses football game and parade were among the new things in 1902.

We are also on the threshold of one of the really startling innovations of the new century—the motor car, as we have seen, and of course, the truck. Horses or mules were still the sole means of heavy transport. Teamsters were for hire, and the company did not have to maintain stables and wagons with all the other paraphernalia needed for operation.

A teamster by the name of "Old Dutch Charlie" served our needs. He lived right across the street at 23rd and Hickory and stabled his horses out in back of his place. Raw materials, however, were received by rail on the spur track that we had and still have at the Hickory plant, and materials were unloaded by stiff-leg crane. This was slow and probably pretty dangerous, but was the only method that we had. Certainly quite a bit of our fabrication or manufacture was likewise shipped out by rail car, especially sizable items. We finally had a truck, probably about 1919 or 1920, and my father remembered it had a tiller, not a steering wheel.

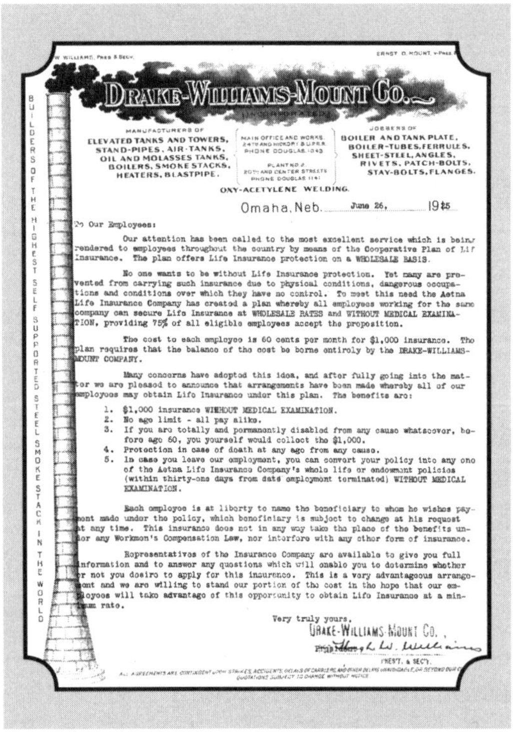

W. WILLIAMS, PRES. & SEC'Y. ERNST D. MOUNT, V-PRES.

DRAKE-WILLIAMS-MOUNT CO.

MANUFACTURERS OF ELEVATED TANKS AND TOWERS, STAND-PIPES, AIR-TANKS, OIL AND MOLASSES TANKS, BOILERS, SMOKE STACKS, HEATERS, BLASTPIPE.

MAIN OFFICE AND WORKS. 24TH AND HICKORY S.U.P.R.R. PHONE DOUGLAS 1043

PLANT NO. 2. 20TH AND CENTER STREETS PHONE DOUGLAS 1141

JOBBERS OF BOILER AND TANK PLATE, BOILER-TUBES, FERRULES, SHEET-STEEL, ANGLES, RIVETS, PATCH-BOLTS, STAY-BOLTS, FLANGES.

OXY-ACETYLENE WELDING.

BUILDERS OF THE HIGHEST SELF SUPPORTED STEEL SMOKE STACK IN THE WORLD

Omaha, Neb. June 26, 1925

To Our Employees:

Our attention has been called to the most excellent service which is being rendered to employees throughout the country by means of the Cooperative Plan of Life Insurance. The plan offers Life Insurance protection on a WHOLESALE BASIS.

No one wants to be without Life Insurance protection. Yet many are prevented from carrying such insurance due to physical conditions, dangerous occupations and conditions over which they have no control. To meet this need the Aetna Life Insurance Company has created a plan whereby all employees working for the same company can secure Life Insurance at WHOLESALE RATES and WITHOUT MEDICAL EXAMINATION, providing 75% of all eligible employees accept the proposition.

The cost to each employee is 60 cents per month for $1,000 insurance. The plan requires that the balance of the cost be borne entirely by the DRAKE-WILLIAMS-MOUNT COMPANY.

Many concerns have adopted this idea, and after fully going into the matter we are pleased to announce that arrangements have been made whereby all of our employees may obtain Life Insurance under this plan. The benefits are:

1. $1,000 insurance WITHOUT MEDICAL EXAMINATION.
2. No age limit - all pay alike.
3. If you are totally and permanently disabled from any cause whatsoever, before age 60, you yourself would collect the $1,000.
4. Protection in case of death at any age from any cause.
5. In case you leave our employment, you can convert your policy into any one of the Aetna Life Insurance Company's whole life or endowment policies (within thirty-one days from date employment terminated) WITHOUT MEDICAL EXAMINATION.

Each employee is at liberty to name the beneficiary to whom he wishes payment made under the policy, which beneficiary is subject to change at his request at any time. This insurance does not in any way take the place of the benefits under any Workmen's Compensation Law, nor interfere with any other form of insurance.

Representatives of the Insurance Company are available to give you full information and to answer any questions which will enable you to determine whether or not you desire to apply for this insurance. This is a very advantageous arrangement and we are willing to stand our portion of the cost in the hope that our employees will take advantage of this opportunity to obtain Life Insurance at a minimum rate.

Very truly yours,
DRAKE-WILLIAMS-MOUNT CO.,
Hugh W. Williams
PRES'T. & SEC'Y.

General letter to employees from Hugh in 1925. Notice the unusual stationery.

We did not have running water at the plant for a number of years. On the payroll (old records refer to simply "kid") were boys who were responsible for fetching water. This was some distance away, down the spur track toward downtown, and was toted back in buckets several times a day to the men. These youngsters were paid about 5 cents per hour. Were we in violation of the child labor laws? Maybe that is why they omitted any names in the payroll records.

Speaking of water, many small towns and municipalities were in need of facilities for developing water distribution for their homes and businesses, and we did a brisk business in water towers and standpipes. This brings to mind a very interesting glimpse that we were afforded into the business and the activities of that particular time.

Cowboys and Indians

Sometime in 1973, I received a very intriguing letter in my mail. It was from the daughter of one Theodore "Teddy" Evert, to whom we are certainly indebted, who took down his words and his remarkable story. It follows:

"An Unexpected Assignment"

Mr. Theodore William Evert was born in Nebraska 6 June 1880. The following is a story he wrote in longhand, late in December of 1972, at his home in Laurel, Montana, to supplement his story told many times to his children and grandchildren.

"Mother and I were married 18 August 1907. I was working for Drake-Williams-Mount Co., contract boiler shop at 24th and Hickory Street, Omaha, Nebraska, building standpipes. Also, American Smelting & Refining Co. Standpipes were one hundred feet tall, and were built at Auburn, Papillion, Benson, South Omaha, Osceola, Stromsberg, Campbell, Lincoln, Minden, Guthrie Center, Dunlap, and a gas holder at Maquoketa, Iowa. At the time we were married, the job was at Lincoln, so mother and I lived with her sister in Lincoln for a while—then moved to Omaha.

In October of 1907 I was sent to Cheyenne River Agency, South Dakota to erect a tank and tower for the government—with two men besides myself. We left Omaha about 2:00 p.m. via railroad. The company always advanced me with sufficient monies in the form of a check. Before leaving Omaha, I stopped at the U.S. National Bank to cash said check, and I was amazed at what I saw in the bank. Monies were stacked high all over the place. I had not ever seen so much before or since then. The reason for all the money I saw in the U.S. National Bank was the beginning of the 1907 money panic. Guess it was the biggest ever. That is why Mount was sending me money in $25 denominations, with instructions to cash them in the post office by buying postage stamps. That panic lasted until late spring, 1908. Script was issued as pay for wages. No jobs—only the ones that the companies could afford, and a lot of men laid off. Mother got by, and did not have any trouble. Neither did I. 1907 sure was a bad year. No money, no jobs.

Well, we left Omaha about 2:00 p.m., by rail, and arrived at Gettysburg, South Dakota some time late that evening, and stayed overnight. The next morning, we left by stage for Forest City, South Dakota, arriving there about 12 noon. Had dinner there, and I noticed a lot of hurrying by men on horseback, and was told about the Indians having Colonel Downs, Indian Agent, surrounded in Thunder Butte Schoolhouse, threatening his life because he had cut off their rations. After dinner, we went to the Agency by ferry boat, and I presented my letter of introduction to Mr. C.W. Rastoll, Assistant Indian Agent. He noticed the return address on it, and immediately we were taken to the store room and outfitted with fur caps, coats, gloves and blankets, and told to get in a lumber wagon with eight or ten other men. Off we went to Thunder Butte with 90 miles before us. We stopped about 6:00 p.m. for a snack and coffee, then off all night toward Thunder Butte, arriving at an Indian schoolhouse at 6:00 in the morning. Had a bite to eat, and coffee. Had to leave our mess wagon there because our horses were so crippled they could go no further. So Mr. Rastoll ordered us to fill our pockets with army hard tack, and we still had about 70 miles to go—where we were heading for at noon. We stopped at an Indian's home, where we had dinner prepared for us, consisting of beef and spuds, cooked in a huge iron kettle, which we enjoyed very much—considering.

After dinner, about 2:00 p.m., we spotted some flashing by mirror, by about six horseback Indians. Rastoll held a conference with us, as to what we should do about that. Presently, they came riding toward us, and we decided to stop them. Asked them what they were doing, and they said they were looking for some horses, and had we seen any? We had not, and they rode down the trail to a stream over to the Moreau river, in to Thunder Butte. We went our way to Thunder Butte, and about there, we sent our scout in to find out how Colonel Downs was. He sent word back to us, "unless we were prepared to fight, not to come closer than seven miles." We sure didn't, and camped for the night. The next morning, C. Rastoll had gone in to see Colonel Downs, and they came back, and upon arrival, they stopped. Colonel Downs got up in the buggy, and said something like this: "Good morning fellow Americans. Glad to see you, but sorry that you came, because you had sent me word that you were not prepared to fight." I, for one, had drawn a single shot gun, and wrong shells of caliber that did not fit the gun. Then he said we could return to Forest City, wishing us farewell. Whereupon we started back, and only about ten miles, lame horses. The second day, about 11:00 a.m., we noticed a terrible cloud of dust ahead of us, and to our surprise, it was caused by the 2nd Cavalry from Des Moines, Iowa, and about 500 Infantry men from Ft. Crook, Omaha. The commissary from Kansas, and Fort Omaha—in all, about 1500 men.

The Ute Indians had a force of about 350, and our outfit of about 30 white men, and about 30 Sioux Indian Police, and 2 blacks. We made about 20 miles back that day, and finally got back to the Agency in 2 more days, and got started at sight of our job, erecting a tank and tower for the government.

The team I still feel sorry for were those dapple greys. They were hitched to a spring

> wagon to go from Forest City to Gettysburg for guns and shells—all they could get—and come to the aid of Colonel Downs. It was about 20 miles from Forest City to Gettysburg. They sure ran those horses, but they stood it, and overtook us, by the time we were camped for the night—a distance of about 130 miles, but they stood it.
>
> The dust stirred up by the Cavalry, with the stars and stripes sure had the Indians frightened, and Colonel Downs escaped their threats.
>
> We finished this tank, and he gave me a letter of acceptance for having done a satisfactory job. Drake, Williams, Mount Co. paid me an extra $20. That Company now is under charge of one Great Grandson of Williams."

As a postscript to this story, I have in my file a picture of a standpipe being completed in Dunlap, Iowa. It is dated 1909, and I would make a small wager that one of the five fearless souls practically hanging from the top of this standpipe is "Teddy" Evert himself. After all, in his own words, he mentions "Dunlap" as one of the jobs that he did.

My father, Arthur Williams, remembered Teddy Evert as a valued employee of the company. I am sorry that we have no record of any of his descendants and thus no one to thank for this engrossing story. I only wish I had been given a chance to meet Teddy—and meet my own grandfather, as well. I have always had a feeling that I would have gotten along famously with them.

Speaking of motor cars, it seems that other than the buggy that Hugh used for his personal transport, when autos did come along, Hugh never adjusted to motorized transportation. For some reason, he never learned to drive. His daughter, Georgia, and my father's older sister drove him everywhere on business trips or on personal

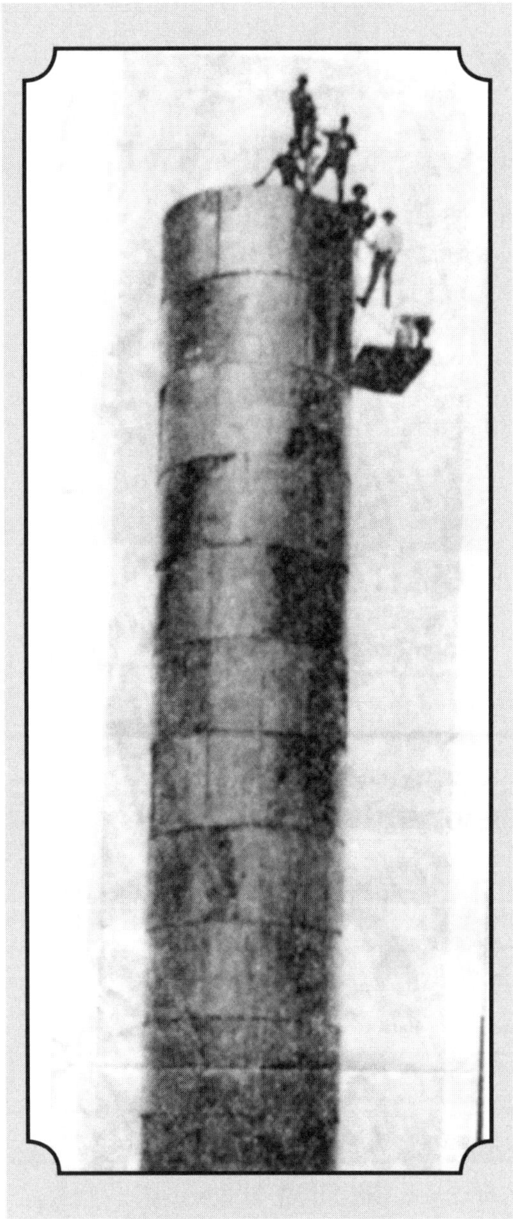

Standpipe at Dunlap, Iowa, 1909. Old Ted Evert was probably one of these "daredevils."

excursions. (This Georgia is not to be confused with Georgia who is my cousin and who is the niece of the Georgia mentioned earlier.)

This meant that Georgia, Hugh's daughter, was privy to much of Hugh's personal and private affairs. Hugh lived many years away from his home with Lizzie. Always supporting her and providing everything that his family needed, he nevertheless lived downtown in the Elks Club Hotel much of the time.

Lizzie and other members of the family did frequent traveling, but he did not accompany them, and seemed little interested in travel. He did make one trip that we know of, however, in November of 1924. This was out West, including to Tijuana in Mexico. We have a postcard from him, sent to the company (no single individual). He says: "Sunday afternoon, Clear and warm, Served real Beers, back Sunday 11/29/24, Hugh."

The reference to "Beers," of course, was because in the states at this time the Prohibition Amendment was in effect, and the only place for obtaining beer was in the illegal speakeasy or from the bootlegger. One wonders if he was by

Hugh sends a postcard home to "the shop" in 1924. Says, "Served real Beers." America under Prohibition at that time.

himself on the trip, but that would be unusual. He could have had a lady friend along, but there is no evidence of this. Prohibition, the "noble experiment," was undertaken in 1920, and was not revoked until 1934. At any rate, it is hard to believe that he was so thirsty that he would actually travel all the way to Tijuana just to be able to legally buy beer.

Some Boom Times

In the years following 1910, and sometime before the World War, business was so good that D-W-M needed more space, so a building and yard on the east side of 20th Street at Center was leased, and some operations were done there.

George Hoody, who came to the company from Colorado Fuel and Iron in Pueblo, Colorado, was hired in about 1910. Hoody was a layout man who later became Foreman, which was the title we used, instead of Superintendent. George told me that at one time, probably about 1919 or 1920, he was laying out work for 150 men. You need to remember that all seams were riveted at this time, as welding was yet to be utilized in the industry That required laying out and center punching every single hole—and there were thousands. Also, those holes needed to "match up" to make a workmanlike fit—no small task.

It was likely that George learned his pattern making and layout while employed at C F and I, but there was another man by the name of Frank Wagner who was probably skilled in the trade as well. George was schooled by Wagner, and George was the man to take over the entire shop force. In the picture of the shop force in front of the old wooden shop, George Hoody can be seen in the middle row on the extreme left. We do not seem to have a better picture of him.

When we consider the life of the shop man, or boilermaker, of this era, it is not difficult to understand why so many hands were needed. The tools and equipment were really limited, and much lifting was done by hand. Outside of the several stiff-leg cranes, about the only lifting was by virtue of hand-operated chain hoists with very limited capacity. Shop work was hard and dirty. All joining was done, as we have seen, by riveting or bolting. This meant every single piece needed to be "laid out" by hand, match marked, center punched (with steel center tool and hammer), and then either drilled or hole punched. Two men were needed to steady the piece and to maneuver it under the punch, just at the right time and place, to be accurately done.

Heavier material had to be cut using a cutting torch or perhaps a splitting (rotary) shear, and again, the materials had to be suspended

Shop force photographed in about 1926. George Hoody, the eventual shop foreman, is in the middle row, far left.

awkwardly from a jib crane and fed into the shear at just the right angle. These operations required special skills. Bucking and driving hot rivets was agonizingly slow and hot, hard work. It guaranteed anyone a really sound sleep that night.

Riveting was not replaced by welding until the early 1930s, and then only gradually. As we have seen, welding was tricky and difficult to master. By 1917, we had about seventy-six shop employees, and into the mid-1920s, that number was up to well over 120, and George Hoody was doing the laying out for all of them.

There is no record of when we pulled out from this property on 20th Street, but it probably was after the war, most likely in the late 1920s. Of course, we know of the Depression and the panic on Wall Street that ensued.

Original stock certificate from AkSarBen Exposition Company.

Hugh invested in local business and alongside of his fellow businessmen in the area. As mentioned, he was one of the original supporters and bought five shares at $100 each of stock in the AkSarBen Exposition Co. The certificate was issued on July 5, 1921, and is signed for the company by E. Buckingham, President. It is certificate no. 1107. In a questionable transaction, the Knights of AkSarBen called in all of the remaining shares, or those that they could find, and issued, in exchange, "memberships" in AkSarBen, in November of 1988.

Also, in September of 1927, he subscribed ($25) to the Air Hanger Fund, which was called "Omaha Legion Airport Corporation" with an authorized capitalization of $30,000 and was "being formed."

1919 was a momentous year in Omaha's history. The World War was barely over, and if that wasn't excitement enough, racial troubles boiled in the city. Plentiful jobs in the packing houses and on the railroad had attracted a lot of in-migration, so there was racial and ethnic friction. Many blacks were attracted by those jobs at this time. There had been a riot in South Omaha just ten years earlier, and the summer of 1919 was hot.

In September, a nineteen-year-old laundry worker, Agnes Loebeck, was raped by a black assailant. The town went crazy, and the police rounded up forty-five black men by the next day, arresting any who could not give an account of himself. One was a man by the name of Will Brown, who reportedly lived with a white woman. He was taken by the mob to Loebeck's house on a Friday evening, and he was nearly lynched when she identified him as her assailant.

The following Sunday, a mob led by Loebeck's boyfriend marched to the courthouse, where Brown was jailed. The crowd grew to huge proportions. The agitators did their job well, and a ten-hour melee ensued. At 10:30 p.m., Mayor Edward Smith brandished a gun and said to the crowd: "If you get him, you have to get me first!"

The crowd attacked the mayor, and he was fitted with a noose and nearly lynched. Police rescued him, but the rioters got to Brown and stripped off his clothes. The mob, in the worst and most infamous chapter in Omaha's history, hanged him from a lamp pole, riddled his body with bullets, and he was tied to a car and dragged through the streets and then strung up on another pole. The riot continued through the wee hours of the morning, and it took troops from Fort Omaha plus the police to quell the disturbance.

Two died besides Brown, both white, and forty-one people were injured. The riot produced a Pulitzer Prize–winning editorial by Harvey Newbranch, editor of the *World-Herald*. Certain prominent Omahans said the riot was provoked, and Major General Leonard Wood (Fort

Leonard Wood, Missouri) who commanded the troops restoring peace, blamed the *Omaha Bee* and the "old criminal gang."

The mayor, Smith, was voted out, and this marked the beginning of the dominance of the Tom "Dynamo" Dennison machine in local politics, which lasted until 1933. The repercussions locally were all about labor, and nothing much about racism.

In our files is a newspaper clipping dated October 23, 1919, from a Walter Johnson. Here is the word-for-word story:

> Johnson's experience with Omaha Police
>
> Oklahoma City, Okl. Oct. 20.- To the editor of the Bee: In an issue of a Nebraska paper I have just read the statement of Commissioner Ringer in his attack on the Bee. In that statement, Ringer says:
>
> "If Mr. Rosewater had told the truth, he would have said that the indirect cause of the riot was the undercurrent of the after-war social restlessness such as follows all great wars, and that the direct cause was the crystallization of mob spirit by vicious, unprincipled, and false newspaper criticism of the police department. These vicious anarchistic attacks by the Bee have been constant for the last eight or nine months. Inspired by newspaper stories, a lynching party was formed at Gibson and marched to the court house."
>
> Was the news of the assault on the woman a story? Did the Bee instigate that so-called story to infuriate the public? Ringer says that the newspaper stories was (sic) the direct cause of the mob. What are the people of Omaha and Nebraska made of, if they were

to believe one word that Ringer or the State Journal would say except their statements were backed by other evidence. Ringer and the city government were the indirect cause of the mob, and the assault on the woman was the direct cause of the mob. Why should Ringer put such a statement out to the public with the expectation that the public would believe it? If the truth was told in the newspapers, the assailant of the woman had confessed to the authorities and that information had been published; the assailant had also been identified by both the victims of the assault. Was it a story or was it the truth? Was the assault of the woman the direct cause of the mob, or was the direct cause of the mob the newspaper stories of the Bee? The people will not be fooled on such trash as the statement that Ringer puts out, and they will not be moved by the preaching of a few weak-minded ministers who are ready to defend so-called law and order. The trouble with Ringer is that he is a natural trickster, and he is overloaded with a case of swell-head, and he believes that what he does not know is not worth knowing.

In the early part of 1918, I had built a heavy piece of machinery, and assembled it on South Twenty-Fourth street, Omaha, and secretly three or four fellows undertook to break my business, and in attempting to do so they came onto my premises and made three separate assaults. I appealed to Mayor Smith and Ringer directly, and Smith told me they expected to elect a police judge some time in the future, and Ringer had another excuse. Eberstein was at that time head of the department of justice, and being that I had certain letters from the government concerning operation of the machine, I appealed to him when the city authorities failed to protect my business. Eberstein was just like Smith and Ringer. As a matter of fact, we lost thousands of dollars because of the anarchists who were holding office in Omaha, who refused to extend police protection, We did not ask for anything except protection from assault. And we did not get it.

WALTER JOHNSON.

Frontier Justice?

Of course, this was not the first racial incidence of lynching, and one which sent a shiver of lawlessness and fear through the town. This account is from the pages of *A Dirty, Wicked Town* by David L. Bristow. This was in 1891 and was fomented by a purported rape, also, of a five-year-old child named Lizzie Yeates. It was determined that the crime was committed by a black man, and Thursday morning, October 8, on page four of the *World-Herald* the headline proclaimed:

HE IS TOO VILE TO LIVE

Every Reason for the Lynching
of a Lustful Young Negro.

A Little Girl Outraged in the Heart of
The City—Room for a Lynching Committee.

Hard to imagine more incendiary language in print, but it was the World-Herald's view that swift and brutal justice was required, of a kind that could not be entrusted to the judicial system! Details were sketchy, but a Negro, George Smith, was cornered in a hay loft a half mile from the Yeates residence and was taken into custody by two Omaha detectives.

It so happened that another prisoner, a white man by the name of Ed Neal, was due to be hanged for murder on Friday, October 9. A gallows had been built by the county jail, at Fourteenth and Davenport. The *World-Herald* opined that the gallows might properly do double duty.

The present-day staff and executives at the *Omaha World-Herald* must truly cringe at the recalling of these words and stories, which would never be tolerated in today's world.

Neal was hanged on schedule at noon that day, but the fate of George Smith was still in the balance. An angry crowd gathered, even supplanted by others who came from Council Bluffs. Efforts were made by officials ranging from the police and firemen to the Governor of the State, James Boyd.

The two papers, the *Herald* and the *Bee*, vying with one another to see which could be the more sensational and outrageous, really fanned the flames of public indignation and encouraged the mobs that would not be satisfied with the slow-moving wheels of justice. They even published unconfirmed reports that the little Yeates girl had died. The mob finally won out, breaking into the heavily fortified jail, and getting to their prey.

Smith was battered and beaten with hundreds of blows. The police made one more attempt to take Smith from the mob, stuffing the broken man into a hack, but the crowd overturned the carriage and smashed it to bits. A noose was fashioned and the rope was thrown over the streetcar wire. He was raised above the crowd, but the hanging was superfluous. Smith was already dead, and his death therefore was not technically on the hands of any individual, but of the multitude.

Mayor Cushing called it the worst thing that had ever happened in the history of the county. Little national attention was paid to the lynching, and, sadly, not much change seemed to have been made in the public's consciousness, certainly not change enough to head off the lynching of 1919. Omaha was indeed a dirty and wicked town. Why, even such a personage as Rudyard Kipling, in 1889, had found Omaha to be "populated entirely by Germans, Poles, Slavs, Hungarians, Croats, Magyars, and all the scum of the Eastern European States."

The Walter Johnson clipping was probably cut out by Hugh or Walt Williams, or Mr. Mount. The story is, of course, of the riot and hanging of Brown. The last part about the heavy machinery on South 24th Street most likely refers to our plant. Perhaps he had a crane and maybe a prototype of one of our dragline buckets, trying to perfect the design for balance. We have no record of a "Walter Johnson" in our files. The original idea (was it copied?) was supposedly brought to us by Fred Crane, who was a dirt moving contractor, and it was about 1913.

Saga of the Omaha Dragline Bucket

There was another dirt contractor that they called "Big Wagon," but his last name is believed to be Peterson. Other early contractors around the Omaha area: George W. Condon; Callahan and Walker; Briggs and Drew; Cole Brothers; and Peterson, Shirley and Gunther.

The "Shirley" was Paul Shirley, and I knew his son, also Paul, but he did not follow in his father's footsteps, but went into the banking business, working for Landen's Security National Bank.

John W. Page, a Chicago engineer, first developed a dragline bucket for excavation and rehandling of dirt and materials, and it was used extensively in the building of the Erie Canal and also the Panama Canal. The equipment was especially efficient at working around water and in tight places where tracked or wheeled equipment had trouble reaching.

Sometime in 1913, Fred M. Crane, a large Omaha dirt contractor, marched into the company's offices with an idea and some sketches for a lighter weight bucket. The big thrust of that era was to increase payloads, and lightening the equipment meant more dirt moved and more production in the same span of time. So, we made the "lightweights" and the designs were pretty good, and we began making the buckets before we ever came up with the name "Omaha." Business was so good, in fact, that the Page people pointed out that they happened to have a 1905 patent on the whole principle of the dragline bucket. After the intercession of the attorneys, a deal was made, and we paid the Page Engineering Co. a patent fee, probably up until the patent expired in about 1922.

That was when we coined the name "Omaha," and the hundreds that were built were worked up and down the Missouri and Mississippi rivers. Omahas were used in World War II on Saipan, Guadalcanal, Okinawa, and on Omaha beach in Normandy—literally all over the world.

The first buckets were all riveted, as was everything else until along about 1920, when electric arc welding arrived on the scene. Arthur Williams remembered the first welding machine that we had, which was a General Electric DC model. The process used a bare wire and lime powder flux, which had to be "dusted" along and was a difficult and tricky job at best. Our first welder was a man named Virgil Burge who had learned the skill at Eaton Metal Products of Omaha and who taught our first apprentice welders the art.

Arthur also remembered that our first large welded job that we tackled was the ductwork and breechings at Phoenix Utilities, at the

Hugh Williams proudly showing a 5-cubic yard Omaha dragline bucket. Notice all riveted construction.

6th and Jones Street plant. This was later to be Nebraska Power Co. and, still later, Omaha Public Power Co.

Two pictures in our files show Hugh posing with a very serious demeanor next to "Omaha" dragline buckets. The company chose to attend a convention of the National Drainage Congress, September 1922. Probably we were anxious to be able to offer our buckets free from the royalty encumbrance, and now boldly called "Omaha." He stands on the left in suit and bow tie, and eldest son Walter is on the right, with a small model of a dragline bucket in his hands.

The other photo is said to be from October 1925, but it looks older than the 1922 picture. Hugh stands with his right hand resting on the trunnion of a 5-cubic yard bucket, which looks like a rugged, tough customer. Again, Hugh is stern in hat and vest, with his clothes looking

An Omaha dragline bucket doing a nice excavating job for a high-rise in downtown Omaha.

pretty wrinkled. He has dark-rimmed glasses and a dour expression on his face. If the dating is correct, he is sixty-three years of age.

As the bucket business developed over the years, the company, in addition to the various sizes offered, also offered four basic types of buckets. These were L-C-R (presumably standing for levee, canal, and river); the Featherweight type; the Standard type; and the Heavy-Duty type. The LCRs were always with perforated bodies and were suited for work in or under water, such as dredging. In our price list of 1957, our heaviest model, the 4 CY Standard, weight was shown as 8,500 pounds and listed for $3,210, or about 37 cents per pound. Probably just the bare steel would cost that much on today's market, and this price included chains and castings, all of which are much higher priced items.

A series of things worked against us in the furtherance of this product. Sometime in the late 1940s perhaps, some new competitors appeared on the scene. The main one was Hendrix in Mansfield, Louisiana. They were ideally situated where much of the levee and canal work was common, and they had another advantage. They had their own foundry and could cast their own manganese steel chain, bucket cutting lips, and other high wear parts. They also probably had a labor rate advantage, so they became firmly entrenched in a short while.

A 100-day strike in 1946 also did nothing to advance our cause. During World War II, when wages and prices were frozen, the United Steelworkers of America began a campaign for representation, and they succeeded in 1945. Arthur Williams said that they took advantage of the situation, telling the men that wages were not really frozen—that the company was using it as an excuse to keep wages down. Art was always bitter about the union and how they succeeded in organizing us. Although it later became the AFL/CIO when the two merged, to him it was always "the damned CIO!"

We had a skeleton office force and were unsophisticated insofar as dealing with union activities was concerned, and we became organized as a result. There is little question that we had no mechanism in place that looked to the welfare of the men, as we were small, with our managers wearing several "hats," and not enough attention being paid to personnel matters. It is no wonder that the USW was able to exploit this situation and to win the election.

When demand slowed, our costs suffered, because we were not able to gang up and build buckets in any quantity, so that product, like the boiler and tank business, began to dry up, but that was in later years to come.

We had other products on the fire from time to time. An old flyer advertised a truck dump box that was an easy addition to the bare truck and chassis that the customer furnished. It was the "Haddon Dump Body" and was easily installed on your Ford chassis by means of "four U-bolts and two machine bolts, which are furnished" ($125.00, F.O.B.

OMAHA PRICE LIST

DRAGLINE BUCKETS

Manufactured by **DRAKE - WILLIAMS - MOUNT CO.** Omaha, Nebr.

NORMAL RATING	1/4	3/8	1/2	5/8	3/4	1	1 1/4	1 1/2	1 3/4	2	2 1/4	2 1/2	2 3/4	3	3 1/4	3 1/2	3 3/4	4
LCR TYPE (REGULARLY FURNISHED WITH PERFORATIONS, CAN BE FURNISHED PLAIN ON SPECIAL ORDER AT NO INCREASE IN LIST PRICE)																		
CU. FT. CAP.	8.01	11.9	14.5	18.4	22.4	30.5	37.	44.5	49.5	58.	64.	71.	77.	84.	91.	98.	104.	110.
WEIGHT COMPLETE	1120	1150	1260	1455	1600	1870	1980	2585	2890	3180	3430	3780	3970	4240	4330	4480	4735	4985
PRICE COMPLETE W/REV. TEETH	$395	$435	$550	$680	$725	$895	$920	$995	$1050	$1165	$1320	$1455	$1590	$1825	$1945	$2065	$2140	$2235
FEATHERWEIGHT TYPE (NOT PERFORATED, CAN BE FURNISHED WITH PERFORATIONS ON SPECIAL ORDER AT NO INCREASE IN LIST PRICE)																		
CU. FT. CAP.			14.5	18.4	22.4	30.5	37.	44.5	49.5	58.	64.	71.	77.	84.	91.	98.	104.	110.
WEIGHT COMPLETE			1450	1720	1820	2300	2600	2900	3200	3650	3750	4300	4600	4700	4800	6000	6350	6550
PRICE COMPLETE W/REV. TEETH			$610	$715	$775	$990	$1035	$1115	$1185	$1295	$1445	$1595	$1750	$1975	$2090	$2210	$2315	$2430
STANDARD TYPE (NOT PERFORATED, CAN BE FURNISHED WITH PERFORATIONS ON SPECIAL ORDER AT NO INCREASE IN LIST PRICE)																		
CU. FT. CAP.		11.9	14.5	18.4	22.4	29.4	35.9	41.5	49.1	56.	63.3	70.7	76.	82.1	88.3	96.1	103.	109.
WEIGHT COMPLETE		1300	1600	1750	1850	2450	2800	3100	3500	4000	4600	5400	5800	6000	6300	7100	7400	8000
PRICE COMPLETE W/REV. TEETH		$465	$675	$750	$830	$1090	$1150	$1235	$1320	$1430	$1570	$1750	$1915	$2125	$2240	$2355	$2490	$2630
HEAVY-DUTY TYPE (NOT PERFORATED, CAN BE FURNISHED WITH PERFORATIONS ON SPECIAL ORDER AT NO INCREASE IN LIST PRICE)																		
CU. FT. CAP.			14.5	18.4	22.4	29.4	35.9	41.5	49.1	56.	63.3	70.7	76.	82.1	88.3	96.1	103.	109.
WEIGHT COMPLETE			1750	1850	2300	2800	3100	3400	3850	4400	5100	5700	6100	6400	6800	7500	7800	8500
PRICE COMPLETE W/REV. TEETH			$820	$925	$1105	$1350	$1420	$1680	$1725	$1885	$1990	$2275	$2425	$2630	$2875	$2980	$3080	$3210

NOTE: All weights and prices shown above are for buckets completely fitted including teeth, ready for operation.

Prices F.O.B. Omaha, Nebr. Effective February 25, 1957.
Supersedes price list dated November 24, 1954.
All prices and specifications shown are subject to change without notice.

For prices on larger or intermediate sizes or special types, contact factory direct.
Weights subject to variance of 5% plus or minus.

1957 price list, Omaha dragline buckets.

Omaha, Neb.). I find no record of how many of these we managed to sell, but apparently we never did much with this one.

We tried to catch on with the booming auto manufacturing business. We made trunks, of luggage carriers for one such manufacturer who was actually located here in Omaha, and who called their car "Moon" or the Moon Auto Co. Several attempts by Omaha start-up aspirants were made to gain entry into the flourishing auto business, perhaps three or four different companies. Not unusual, considering that upwards of some 400 separate aspiring entrepreneurs nationwide were involved at various times.

Pretty slim odds, but Omaha apparently lacked many of the necessary components to make a serious run. Few tool and die makers, machinists, and other skilled people is one example. Venture capital was lacking. Engineering talent and other skills were available in Detroit and Akron and Lansing, for instance, but not in Omaha.

By now, the new location is firmly established, however, there on Hickory Street and apparently enjoying thriving business. Also by now, Hugh has one more mouth to feed—Arthur W., my father, born May 12, 1904, and to be the last child. He is indeed the baby in the family, there being nine years difference in age between Arthur and Georgia, and fully twelve years between Art and Walter.

Unfortunately for Arthur, Hugh removed himself from the Williams household, probably about ten or eleven years later. Not able to get along with Lizzie apparently, but never neglecting to provide generous financial support, it seems that he never asked for a divorce at any time. These were important years for a young boy, growing up without a father to relate to. It prompted Art to complain in later years that he had to grow up with three "mothers" in his family. Hugh made his home in the Elks Club Hotel downtown for a number of years, probably until sometime in 1929.

According to some stories, Hugh did not want for the company of other females. One story that has been passed down concerns a very real threat to Hugh of bodily harm. Word somehow came that the husband of a certain lady friend of Hugh's was returning from being away and was headed for his home, supposedly with a gun, but where Hugh was in enjoyment at this particular time. Georgia, who was his "driver" since Hugh never learned to drive his own car, was sent speeding to rescue him from this dire situation, and did.

A product that was to contribute to the war effort in about 1916 was related to the potash mines in western Nebraska. The potash was a vital ingredient for explosives, and we built dryers and pulverizers for processing of these materials. Electricity was not available out in the hinterlands as it is today, and the dryers had to be steam driven—a fact that led to even more business for the plant. This business faded, of course, after those war years, so the bucket business was a very timely substitute.

The dryers and associated equipment was just another product, and an example of what evolves from a "job" shop. That this company, through the years, has always been reactive rather than proactive cannot be denied. Many ideas, some ours, but most often from customers proved again the old adage about necessity being the mother of invention. But our thrust always remained very general in nature, a characteristic that has stayed with the company throughout the years. Always, the choice was against specialization and toward generalization. Of course, our equipment has always been in that direction and seldom single-purpose.

As we saw earlier, little villages and towns were growing larger and needing facilities of their own. The company not only fabricated a lot of standpipes and water towers, but they sent erection crews west to many points—even as far away as New Mexico and Montana. These early contracts were apparently estimated and drawn up by Hugh and had to be very complex, with all the travel and board involved. At least, there is no further mention of interference or problems with Indians as encountered in South Dakota by Teddy Evert.

The Smokestacks

Drake-Williams-Mount also built smokestacks. There were two large stacks at Lead, South Dakota. The company stationery of the day boasted of the "largest self-supporting smokestack in the world." This was the 311-foot stack built for the Omaha Smelter (later American Smelting & Refining Co.), which was the tallest at that time. It was

23 feet in diameter at its base, and 12 feet in diameter at the top. In the early 1940s, 100 feet was removed from the stack, after concerns for the structural stability were expressed. To the contrary, the material showed no corrosion or deterioration in any measurable amount.

Other tall stacks were built: a 250-foot stack for the old Nebraska Power Co., and a similar 250-foot stack for the South Omaha Sewage Treatment Plant in later years. The big difference was, the stack for both the Omaha Smelter and Nebraska Power were entirely riveted construction, and the stack for the South Omaha Plant was welded.

Large smokestack for Nebraska Power Co. (later OPPD), 1911.

That we were very proud of the stacks we had designed and built was evidenced by our letterhead stationery of the day. Up the entire lefthand margin of the sheets was a representation of a tall stack, smoke issuing out of the top, and forming the background for the bold "-DRAKE-WILLIAMS-MOUNT CO.-"

One particularly good example is an actual letter sent to Arthur Williams, from Walter, his brother. However, instead of featuring the stacks, it concerned our buckets. The tone of the letter seemed to indicate that we were in a pretty good competitive position and were deciding whom to "favor" with a carload of our product. Whether they would land there on consignment or not is not clear. Also interesting is the date of the letter, February 27, 1929—just about eight months before the stock market was due to take a giant nose dive on Wall Street. Art was sent the letter in Memphis, where he evidently was spending a lot of time calling on machinery dealers in the area.

Arthur spent quite a bit of time in Memphis, in Arkansas, and Louisiana—all areas where a lot of river work was going on, mostly on the Mississippi, and some also on the Missouri. He spoke to me about the wonderful seafood that he enjoyed—jambalaya, crawdads, and shrimp that were all fresh and that the locals really knew how to prepare. We had a product that was readily accepted but that later was largely grabbed off by Hendrix of Mansfield, Louisiana, as I pointed out earlier.

Not much is available or was saved of Hugh's writings, but we have examples of his handwriting in the form of his signature, which was very even and attractive, almost feminine in appearance. We have one letter, directed to employees, which may have been authored by Hugh or else could have been prepared by the Aetna Life Insurance Co., which was the company offering a plan of insurance for the men. The plan could be among the first group insurance programs to be offered to businesses. They may have written the letter and put it on our letterhead for his signature, or it could be his own words. It is on the same letterhead with the smokestack rising along the lefthand margin,

It provided for $1,000 worth of life insurance, no evidence of insurability, no age limit, a full disability feature, and 60 cents per month, all paying alike. The balance of the cost would be borne by the Drake-Williams-Mount Co.

Hugh signed the letter as president and secretary, which is not legal these days under corporate law, but perhaps was at that time. The letterhead also indicated "Plant no. 2, 20th and Center Sts., phone Douglas 1141." Also, it boasted of Oxy-Acetylene Welding, and this was dated June 26, 1925. This, of course, was not the same as electric arc welding and was a submerged arc process and difficult to master. It was quite a while before we began to replace rivets as the main method of joining iron and steel members.

Lizzie (Hugh's wife), Hugh, young Janet and mother Janet, about 1930.

Hugh Williams Dies

Hugh and Lizzie made their home in the red brick at 505 S. 51st Avenue. Hugh had the home built sometime in the mid-1920s. He instructed the builder that he wanted a more or less exact copy of a home in another part of town. Later, Uncle Walter, Art's brother, built a home almost identical to this one as well! It was here that Hugh finally came home to, probably sometime in early 1930. He was sick for nearly two years before his death, August 3, 1931.

Again, according to my cousin Georgia, Hugh had my parents bring me in to see him when he was on his death bed, and I was not even a year old. She said he was very pleased, since I had been named for him, and he had only one other grandson, which was my cousin Bud, Walter's boy. My mother told me the story also, and that he avowed that I was a "fine boy."

At Drake-Williams-Mount Co., this left Ernst Mount in charge, and he was president of the company until his death in 1934. With Mount only president for about three years, this now left the company in the hands of the Williams brothers, Walter and Arthur. Walter was about forty-two. Arthur, twelve years his junior at thirty, was a young vice-president and secretary.

The nation was still in the lingering recession that followed the Depression years of 1929 and 1930. These conditions, plus the disparity in ages of the two men may have created great strain and tension. In fact, about the only thing that the two brothers agreed upon was that they hated Democrats, and that included Franklin Delano

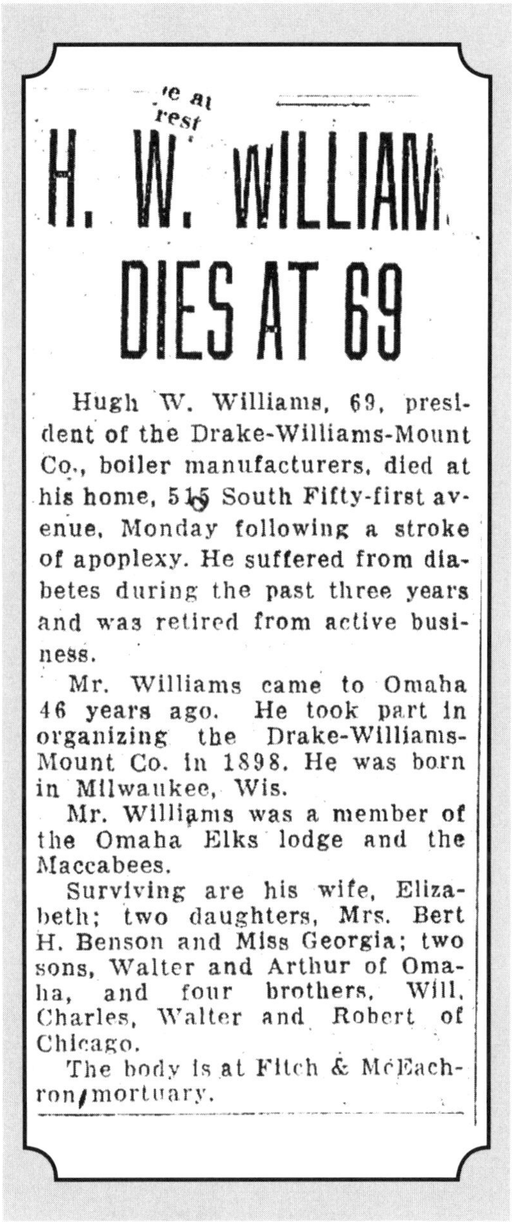

H. W. WILLIAM

DIES AT 69

Hugh W. Williams, 69, president of the Drake-Williams-Mount Co., boiler manufacturers, died at his home, 515 South Fifty-first avenue, Monday following a stroke of apoplexy. He suffered from diabetes during the past three years and was retired from active business.

Mr. Williams came to Omaha 46 years ago. He took part in organizing the Drake-Williams-Mount Co. in 1898. He was born in Milwaukee, Wis.

Mr. Williams was a member of the Omaha Elks lodge and the Maccabees.

Surviving are his wife, Elizabeth; two daughters, Mrs. Bert H. Benson and Miss Georgia; two sons, Walter and Arthur of Omaha, and four brothers, Will, Charles, Walter and Robert of Chicago.

The body is at Fitch & McEachron mortuary.

Hugh Williams's newspaper obituary, 1931

Roosevelt. Sadly, the two just did not see eye-to-eye in business matters, and the cooperation that should have been an asset to the still-young company and its young managers was missing, as they faced what I have termed Second Generation, Depression, and World War.

TWO

SECOND GENERATION, DEPRESSION, AND WORLD WAR: 1932 TO 1956

The Great Depression

After Hugh died in 1931, Ernst Mount became president of the company. Mount's term as president was only three years, as he died in 1934. He was a no-nonsense, hard-headed leader and businessman. He was no doubt just what was needed after Hugh's death and a stabilizing influence for the young Williams brothers. But suddenly, in 1934, he was gone and they were in full charge, and business conditions were not ideal.

The country was in the third year of one of the worst depressions in U.S. history. Politics all over the world were topsy-turvy. For example, in England, Ramsay MacDonald, founder of the Labour Party, was elected to head up the Parliament, which was dominated by the largest Conservative majority ever.

Here, ironically Franklin D. Roosevelt, a patrician, ran and won the presidency based on being a champion of the working class. Hoover, although known as a progressive and a humanitarian, was tossed out. He became known as a reactionary, and one who had done nothing for the nation's impoverished masses. He retains that image even today, despite his establishment of the RFC, the Reconstruction Finance Corporation, which aided financial institutions and the railroads. The idea was to provide businesses with capital, that they may start hiring again. Despite this help, many banks failed, and the people described the RFC as the "breadline for business." Roosevelt campaigned and won with his promise of a New Deal for the forgotten man, and he was elected by a landslide.

Nevertheless, Roosevelt never achieved a degree of prosperity for the country until the hostilities in Europe summoned us to begin producing munitions, and most of the rebuilding of the services was done with deficit financing.

More ominous but not appreciated immediately, elections in Germany spelled the end of democracy for that country. The Nazi party emerged as the Reichstag's largest, and would soon name Adolph Hitler chancellor. Nazi storm troopers battled Communist paramilitary groups almost nightly in the streets. Little was known by the average person about him, despite his writings in *Mein Kampf*, nor of what was to be coming in the future for the German republic and the world.

As if these developments were not enough, in India, Mahatma Gandhi and Nehru, along with thousands of others, had been jailed as that nation used civil disobedience to the British crown in a drive toward national sovereignty. Gandhi began his famous "perpetual fast unto death," which focused world attention on the struggle. Japan, a growing military power, had occupied Manchuria, and now attacked Shanghai.

This action caused a new unity among Britain, France, the Soviet Union, and the United States in opposition to this expansionism. Mussolini was in power in Italy, Stalin in Russia. The stage for world war was truly set (excerpted from *Our Times: The Illustrated History of the Twentieth Century*).

Born in 1932 was Jacques Chirac, French President even as these words are written; Mario Cuomo, Governor of New York, and Edward Kennedy, Senator of Massachusetts. Also, actress Elizabeth Taylor, writer John Updike, and French filmmaker Francois Truffaut. George Eastman, noted U.S. inventor died, as did John Phillip Sousa, the composer known as the "March King."

Along came the "Zippo" lighter, Revlon, the Three Musketeers candy bar, and the first gasoline tax. The world was spinning toward an age of totalitarianism, and many were the unknown dangers that lay ahead.

It was said that unemployment ranged around the 30 percent level, and our small company was just trying to survive, one supposes. The nation enacted the Smoot-Hawley tariff act, raising already high tariffs on imported

Walter Williams, brother to Arthur, president 1934 to 1956.

goods to about 40 percent. Economists pleaded with Hoover to veto the bill, but he felt that foreign competition was unfairly priced, and he would not veto. This was possibly the greatest mistake that Hoover could make, thus ensuring continuing depression-like conditions. The economists were proved to be correct, and the law only made the situation worse. There was a lot to be learned about international economics, it seemed.

We were a boiler repair shop, a tank maker, heavy sheet iron fabricator, and we were manufacturers of several products: "St. Paul" hydraulic dump bodies, "Omaha" dragline buckets, plus straddle tanks, ice brine tanks, and smokestacks and breechings "to order." Presumably, we made the dump bodies under agreement with "St. Paul" and possibly stocked some of their parts for resale. We did no business with building contractors, only dirt contractors. We knew nothing about structural work.

An old tear sheet from this era outlines the many advantages of the "Haddon" dump body and includes a photo of the contraption. "It can be installed on a Ford chassis with four U-bolts and two machine bolts (furnished) with ordinary tools, and will take only fifteen minutes to complete the job. All this for $125.00 F. O. B. Omaha, Neb."

No record exists of how low employment went in these years. Our bucket business was fairly steady, and since steam power was only partly diminished, we kept body and soul together, and even made a little money.

First Ventures in Warehouse Market

We also did a certain amount of "warehouse" steel business in the 1930s. In 1937, we published and distributed a limited catalog or "stock list." It was a spiral-bound booklet, about 5" x 9" with a two-color cover. It was entitled "STEEL IN STOCK"—1937 Stock List. Telephones were listed for the office, JA (Jackson) 1043 and after business hours GL (Glendale) 2098 and WA (Walnut) 5358. The latter two had to be home numbers for Walt and Art Williams. It was a nice little catalog of forty-two pages, about a dozen of which were steel items. We advertised ASME code pressure vessels, flanged and dished tank heads, manhole and handhole fittings and plates, machine bolts, BBB coil chain, wire rope, and many other specialty items.

The warehouse business was the place where Art wanted to concentrate, and he felt that there were opportunities for growth here. Walt did not feel the same way, and having been exposed for so many years to the steam boiler and associated product business, he fixated on this business. That the era of steam was ebbing did not seem to ever occur to Walter, but Art knew that changes were needed and that we needed new directions.

I remember hearing that we hired a salesman by the name of Steve O'Donnel, and I suppose it was in order

Drake-Williams-Mount Co. steel handbook.

to try to garner warehouse steel business. I was very young, and I only saw him once. I don't think he was with the company too long, so this venture was evidently not a big success. I have no idea where he came from, nor where he went, but he was not long with the company.

Acrimony

These were years in which agreement and cooperation were not to be the case between the brothers and principals of the business, Walter and Arthur. They both had lived through a complete business cycle—booming times in the late 1920s followed by a bone-rattling crash in 1929 and 1930. Walter seemed especially affected and averse to any kind of business risk. He often resorted to his motto: "Bricks and mortar!" I was never sure what that was supposed to mean, but I know that, to him, it meant to just keep your powder dry, husband your cash, and don't spend it on anything that is not absolutely necessary! And, we didn't.

These were tough years especially in terms of the relations between Art and Walt. According to Walt, he was the driving force that kept the shop "on the stick" and that took care of engineering and technical matters, and Art just coasted along in the office. By Art's version, "Whoop and Holler," as he was called, was out in the shop, grabbing tools out of the men's hands when he didn't think they were doing something or other "right," and generally getting in the way.

Walt would spend days and days drawing up a smokestack and showing every single rivet and bolt in the process. Then, he would painstakingly write out a proposal in minute detail, after which he would give it to Virginia (Huntington) for typing. She would have a terrible time reading all of his "hen scratches." Then to make things worse, Walt was losing his hearing at a rapid rate, and this was not helping his disposition. All of these years, he never seemed to want to bring his son, Bud, out of the shop to give him a grounding in that aspect of the business itself. Perhaps it was Bud who resisted the change and was simply satisfied to be a welder in the shop.

The truth probably fell about in the middle, and there really was very little contact in the form of sales effort with the outside world. A Williams trait seems to be that, if we know something to be a fact, then we assume that everyone on the planet knows this as a factual matter as well. Therefore, don't waste your time discussing it any further.

There was one thing that the brothers agreed upon, and wholeheartedly so, and that was their hatred of Franklin Delano Roosevelt. They hated his New Deal and his make-work programs, such as the W.P.A. and the CCC (Civilian Conservation Corps). Hitler was obviously up to no good in Europe, and they felt that FDR, the globalist, would drag us into an unwanted second World War, despite his promises to the contrary. I remember Art mimicking Roosevelt when he stated, "Eleanor hates 'wah,' I hate 'wah,' and even Fala hates 'wah!'" Fala, of course, was FDR's little black Scotty dog.

I remember seeing FDR cruising west on Leavenworth Street at the top of 56th Street, which was where we lived at the time. My sister Janet took me up to see the motorcade, which had to be in 1935 when he was campaigning for the Presidency.

Fay Watson Changes the Landscape

In the years that preceded the Watson Brothers purchase of the site just north of us, on 24th Street, the ground directly across the street was more or less level with our property, and there were two or three old frame houses there. The land just west of us, which was about two and a half acres, was owned by Socony Mobil Oil Co., and was a barrel cleaning facility.

Art and the rest of our office staff didn't pay much attention to the older people who lived in these homes, but one day, the milkman who serviced the needs of one of the homes came over to our office and asked if we had someone that would accompany him to check up on the people there. It seems that they had not taken in their usual bottles of milk that he regularly left them, and he was growing fearful that something could be the matter with them.

So they knocked at the door, and, getting no response, they tried the door, and it was open. They discovered the old man and his wife inside, deceased, and no one knew the first thing about relatives (if any) or about whom to contact to report what had happened. They called the police. Art was pretty shook up by the situation, and I could sense it. He really didn't want to talk that much about it.

Not long after this, Fay Watson, who was the president of Watson Brothers Transportation Co., contacted Art and told him he had purchased these lots, as well as about four acres of additional land, and intended to build a terminal building on the property. He explained that he had to bring in fill dirt to make the whole property level, and that this would result in quite a hill, or bank, all along Hickory Street to 23rd Street, and beyond. He assured Art that he would construct a retaining wall and that he would maintain it to prevent any erosion that would affect us on our side. He really laid it on thick, because he feared that we could oppose the idea. That it would have an effect on drainage, there was no question. It did, and it just meant more water than ever would flow into our grounds, and, of course, we have always had flooding problems there. However, Fay, while he was alive, kept his word about the maintenance of that bank, and he kept the weeds down as well.

Meanwhile, in Russia, Stalin began his great purge, but in the U.S., very little attention was paid to this criminal activity. At least equivalent to the later Holocaust in Germany, Stalin systematically destroyed the old Bolshevik Party. In a three-year period that lasted through 1938, with his secret police he arrested 5 million citizens, many of whom were executed. The rest were sent to the Gulags, which were forced labor camps in Siberia.

The United States, desperate for allies to oppose Hitler and fascism, was ready to believe Stalin's lies. On another front, Japan had invaded China, and likewise, this was not getting a lot of attention in this country. I remember a few newsreels that were run at some of our movie theaters, where some of the atrocities being committed by the Japanese were shown. The "Rape of Nanking," one of the worst chapters in the war, was never mentioned. The United States seemed determined to "whistle in the dark" perhaps in hopes that this war business would just blow over and go away.

World War II

"You remember that … it was in all the papers!" (said the comedians). Yes, we certainly do remember!

As everyone also knows, Japan made the decision for the U.S. with its attack on Pearl Harbor. Germany had attacked Poland, and true to its promise, England declared war on Germany. Since Germany had formed an alliance with Japan, we declared war on Germany as well as on Japan.

The later years of the 1930s had gradually improved, but it wasn't until the war years of 1941 and forward that business boomed. It boomed for lots of others, but not so much for Drake-Williams-Mount Co. Our competitors in Omaha got a lot of defense work, but we had a small force of people and apparently did not qualify for any of the contracts that others were nailing down. We probably did not really make a very strong attempt to win some of that war work, and so we lost out on things like equipment and tools that were available and that our competitor friends had handed to them—FREE!

Omaha Steel built barges and landing craft. They built a shop of sorts right on the river, under the South Omaha Bridge, where the craft were built and floated out for testing in the river. Gate City Steel built assemblages for tanks and armored vehicles, as did Paxton-Vierling Steel. Some made shell casings and other miscellaneous items.

We did do some subcontract work under Baker Ice Machine Co. on rear gun mounts for several models of our World War II bombers—Flying Fortresses, Liberators, Marauders, and others. For our war work, the company was awarded an "E" flag, which was awarded to thousands of industries for defense-related work all during that war. Meanwhile, others were having their whole plants retooled and equipped. Not having a flag pole, D-W-M never got a chance to fly that flag!

We made lots of buckets, largely for the Navy. Art traveled to Port Hueneme, California, to pick up several major orders for shipment to the South Pacific and elsewhere. Apparently, we had trouble finding manpower and welders to fill those orders. Selling prices were "frozen," so it is doubtful that we were really getting enough money for these buckets, but it was business, and we needed it.

The B-29, the Enola Gay, piloted by Col. Paul Tibbets and that carried the atomic bomb that struck Hiroshima and brought an end to the war against Japan, was built right here at the Martin bomber plant in Bellevue.

One bright spot that was a good profit center for us was the boiler business. We did a brisk business in boiler repair. There was really only one other competitor in town, C.G. Johnson Boiler Co. That company was owned and operated by Louie Johnson. I never had the pleasure of making his acquaintance, but I do know that we always fought like crazy with him over the repair jobs that were out there. Walt fought with Louie, and they always tried to cut each others' throats for the work. We had a fairly broad stock of boiler and condenser tubes, but Johnson did not buy our tubes.

When I first joined the company full-time, I was sent out on some of these calls, to take a look at these boilers. Naturally, having had no experience, and certainly never having done any of this work myself, I was not much more than an interested spectator. My summary was always the same. This looks really bad, and it is, of course, impossible to know just how bad until we "get in there, and tear things down." Naturally, this always meant that we would not be able to quote a firm price for the work. But, this was me, and we always "got the job."

I imagine that Walt, when he was called out, took along some tools on the call and beat on the tube sheet and tubes with a hammer or wrench and shined his flashlight all around, and then went back to the office and put together a firm bid for the prospective job. We always gave those customers a good job, and always at a fair price. I reasoned that the boiler was not going anywhere, that they certainly needed it for heat and hot water, so they really had no other options anyway. We treated these people right and always charged reasonable prices for our work. We did not gouge, but also we certainly did not lose any money. As a matter of fact, our customers were getting a very good deal, because, unknowingly, we simply did not charge enough to cover all of our overheads, plus a reasonable profit.

We had a good inventory of boiler and condenser tubes, but I'm sure that Johnson never bought from us, unless there was a dire emergency. After 1956, we figured there was really no future in beating up on each other, so we made a deal with C.G. Johnson, which by this time was under new ownership and direction, since old Louie had finally died. We would only do repair jobs that were just handed to us—jobs that were strictly time and material. Other inquiries we would politely decline, leaving the field open to Hank Steele, the new owner of C.G. Johnson. In exchange, they bought most of their tubes from us. We were running short on boilermakers anyway, and we sure weren't training any new ones! This worked out great, since we were becoming busier all the time with our own in-shop work.

We did make a lot of buckets and parts. These were shipped to Europe and to the islands of the South Seas. I think the Navy ordered enough parts that they could have used them for missiles, firing them at the enemy, never running out. I remember we needed hundreds of wooden boxes, made to just exact sizes, to ship lots of these parts. Art decided he would save some money and build them himself in his small workshop, which he had in our basement at home.

Every night, for quite a while, he would go down to his work room, and he enlisted Mike's help. Mike was only about nine years old, but he learned to put the pieces in a jig and nail them snugly together. Art cut up all the pieces, and then they both did the nailing. It was quite a production line.

World-Herald Story

From time to time, the *Omaha World-Herald* would do a feature story, as they did on local business and industry, and in 1945 it seems it was our turn. It was a sizable article, entitled "The Saga of the Omaha Bucket." Also included were pictures of Walter and Arthur and of a large smokestack at the Nebraska Power Co.

AFTER Pearl Harbor, the Omaha bucket was on the job cleaning up the mess. It was in New Guinea helping our troops fight and climb up the Owen Stanley Mountains; it was an early bird on Guadalcanal, Saipan, Iwo Jima, Okinawa, Omaha Beach in Normndy.

Walt Williams

As V-E Day neared, there came through an order for a whole flock of Omaha buckets to help clear the port of Amsterdam. And it's a safe bet that the Omaha bucket will be a member of the assault force that lands on Nagoya to make the final drive on Tokyo.

The versatile Omaha bucket is no secret weapon. It is nothing new. It has been manufactured at Omaha since 1910. Its builders — to quote them — "didn't even have sense enough to get a patent."

Art Williams

The Omaha bucket is just the best darned handler of earth, rocks, coral, sand or debris in the dragline business.

You know—the better mousetrap story all over again. Except that these are pretty big "traps." They trap up to 13 cubic yards of material in a single operation of the winch.

World-Herald *salute to the "Omaha" bucket, 1945.*

Here is the lead paragraph: "After Pearl Harbor, the Omaha bucket was on the job cleaning up the mess. It was in New Guinea helping our troops fight and climb up the Owen Stanley Mountains; it was an early bird on Guadalcanal, Saipan, Iwo Jima, Okinawa, and Omaha Beach in Normandy."

The article is very flattering to us, calling it "the best darned handler of earth, rocks, coral, sand or debris in the dragline business." Further, in connection with the building of the Erie Canal, "the dragline was evolved by John W. Page, a Chicago engineer." "The Omaha names in the field were famous the nation over—Callahan-Walker, Briggs and Drew, Fred M. Crane, Cole Brothers, George M. Condon, and Peterson, Shirley and Gunther to name a few. So famous were they that they did big jobs on the Panama Canal, the All American Canal to the Imperial Valley, and a host of other big earth moving operations."

Fred Crane is credited with spurring us on to come up with a lightweight bucket—one that was lighter than that being made by Page in Chicago. This was about 1913, and when we began making inroads, Page brought suit under a "basic patent of 1905." A compromise was made, and for a number of years, Page and Drake-Williams-Mount alone could build the buckets. Patents have long since run out.

The story goes that once, when Walter and his father, Hugh, were showing off the product at a convention, Walt was asked for the name of the equipment. Walt stepped up, taking a brush and some white paint, and lettered *Omaha* crudely on the arch, and it was forever after called the "Omaha" dragline bucket.

The newspaper further talks about the founding of the company by Hugh Williams, and states that Hugh came to Omaha from Chicago with A.J. Vierling and was "in charge of sales for the Paxton and Vierling Iron Works."

If this statement is true, it had to come from Walter, and being twelve years Arthur's senior, certainly could be true. It would explain why Hugh went to work for Paxton and Vierling and why he did not go to the fledgling Union Stockyards for employment, since he had a lot of experience with that same company as a hog buyer while he was in Chicago. In fact, it also could be the whole reason for Hugh making his decision to come to Omaha.

Hugh and son Walter, with a "low arch" Omaha bucket at an industrial exhibit in Kansas City in 1922. Walter holds a small model of an Omaha bucket in his hands.

Another item in the newspaper story that is wonderful irony: "The firm stuck to its last, keeping out of the steel construction field." No one apparently saw the irony of that statement—and what a difference that could have made!

"Drake-Williams-Mount built two large stacks for smelters at Lead, South Dakota. It built the 311-foot steel stack for the Omaha Smelter (ASARCO), in its day the tallest self-supported steel stack in the world. It was 23 feet in diameter at the base, 12 feet at the top. The top hundred feet has been removed in recent years. Drake-Williams-Mount built the 250-foot Nebraska Power Company (forerunner to OPPD) stack at its South Omaha plant, tallest in these parts nowadays.

"It was right at the time of World War I and immediately following that Drake-Williams-Mount had its largest payroll. Much of the machinery for the Nebraska potash fields was fabricated in the Omaha shops. At the same time the sugar beet industry was coming into its own in the Platte Valley, due to opening of irrigation in the Scottsbluff vicinity, and there was a wide demand for machinery for the beet factories.

"In the potash field business, Drake-Williams-Mount made a lot of money, but wound up by taking a licking when the bubble burst. As the sugar factories were completed, orders in that field fell off. It was time to retrench.

"The firm got out of the boiler making business, but continued in the heavy plate and sheet metal field, making oil tanks, air tanks, water tanks, smoke stacks, and, of course, the Omaha bucket."

The article quotes Walt, on the subject of employment: "The payroll has been kept about the same, 65 to 70, although we could have used a hundred more men if we could have found them." Walt himself claimed to have started at age fifteen for 5 cents an hour, working ten-hour days and six days a week.

"Mr. Drake had died in 1905 [actually 1902], and Hugh Williams purchased his interest. (Incidentally, the money the Williams' paid the George Drake estate figured largely in the construction of the Drake Court Apartments by George Drake's son, William Drake)."

The feature article used eight pictures, including portraits of Walter and Arthur, and one of a thirteen-cubic-yard bucket—so large that a man was able to stand in it.

In the years following, notices in the press were not noteworthy: "2 Steel Firms Hit by Strike" (April 29, 1946); "Union Is Choice of Boiler Workers" (1945); "Long Steel Firm Strike Near End" (August 6, 1946); "Firm Burglarized" (March 20, 1949); "Iowa Arrest 2 in Robbery" (March 27, 1949); "Vandals Drop Crane Boom" (April 20, 1962).

During the war, naturally everything was subject to rationing. Gasoline, sugar, bacon and meat, tires—nearly everything. In business, it was not a matter of ration stamps, but of "priorities," which were simply licenses to buy various commodities, and, of course, steel was one of them. We were not aggressive in pursuing defense work to the extent that others were and, as a result, did not get in on a lot of the profits that were there for the taking.

One of the survivors of the war was the versatile jeep. Used by all the services all over the world, it was to become popular as well at home for all sorts of jobs and leisure activities, too. I was still in school, right after the war, and very few had yet to be seen in the civilian world. But Ajon Farber, a friend from high school days and who was from a wealthy insurance family, suddenly showed up with a jeep. It was a regular Army issue, painted khaki color, and he got in a bunch of trouble for driving it right up the west stairs of Central High School!

A company that did rate the work was given steel, and there was always more than was directly needed for a given contract. This meant that "accidentally" one might have some very scarce steel for certain domestic uses and extraordinary jobs. One of the wartime programs by Roosevelt was "Lend-Lease," which was designed to pour much needed kinds of weaponry, food, and materiel to our allies, Britain and Russia, without payment up front, or even a payment schedule. This came under much criticism, but it probably helped end the long economic depression. It also probably set the pattern for deficit spending that has taken place ever since.

Walter's only son, Bud, was serving in the Army during this war. I have a "V-mail" addressed to Mr. and Mrs. Arthur Williams and Family, and it was a thank you note for a war bond they had bought for Bud for Christmas. (A V-mail was a letter which was photographed and reduced to save shipping space ostensibly, but more likely was easier, as a result, to censor for national security purposes.) Bud wrote that he had been in the Army for more than three years, and nearly three years overseas. It was apparent that he was bored and longing for the day he could get home again.

It was later, sometime in the mid 1980s, when Mother called me to report she had seen in the obits the sad fact that Bud had died. It was too late to attend the funeral, and I have no idea as to cause of death. I think his life was pretty well ruined by Walt, his own father.

D-W-M Unionized

During this time, the nation lived under wage and price controls also. Both wages and prices, according to certain formulas, were frozen, to avoid runaway inflation and to maintain order. It was this very situation that Arthur Williams blamed for the fact that we were organized by the CIO union right after the war, by an election held at our plant. He claimed that the union organizers had advised the rank and file that the company had used the controls as an excuse and that they actually could have given the men a raise. Art, of course, claimed the opposite was true and that we were prevented by the law.

There was a wartime "Wage and Hours Act," which froze wages, the same as the Office of Price Administration had frozen prices on virtually everything. Which version is correct is hard to say, but I doubt that much effort was used to try to raise wages for the men at that time.

One of the welders and mechanics was a guy named John Kiper. John was very active among those who were trying to form the union in our shop. John left after a few years and showed he was an enterprising sort by starting up his own business, Midwest Manufacturing Co., not directly in competition, but in miscellaneous manufacturing and fabrication.

He was located on Park Avenue, next to Olson Bros. Roofing and Sheet Metal. Then he moved way out on North 30th Street, and later, he bought the old building on North 11th Street, which was the former home of Puritan Manufacturing. Many years later, I called on him as a potential customer and had a few talks with him. How he had changed as regards his feelings toward unions! Now, he wanted no part of a union in his own operation and made no bones about it. Interesting to see how one's attitude and mindset changes.

John Kiper was very good to me, taking me on one occasion on a trip to his relatives in Sabetha, Kansas. I was, in fact, visiting him at his small acreage where he lived with his wife, Golda, on the very Sunday when the Japanese attacked Pearl Harbor. I was only ten years old, and I doubt that I understood the gravity of the situation. I'm sure that John's idea was to curry favor with "the boss" when he took me different places, but I think he genuinely liked me and spent a lot of time with me.

It was sometime in 1943 that the election was held. Supervised by the NLRB (National Labor Relations Board), the vote was twenty-eight for the USW, and two for the Boilermakers International Union local no. 38, and three employees who did not vote, and only two who voted for no union. All was legal and proper, but everything changed for the company.

The *World-Herald* story in the issue of April 29, 1946, told of 230 workers going out on strike. Most of them were the employees of Eaton Metal Products Co., and only about thirty were ours. Stover, the representative of the United Steelworkers, said that the union was seeking an 18-and-one-half-cent hourly increase, the same as was awarded to workers in the basic steel industry to settle a nationwide strike. It finally ended after the 100 days with the union accepting a 16-cents-per-hour raise. I imagine that wages were nearly all of the dispute, and there is no evidence to the contrary.

On August 7, 1946, Arthur issued a form letter to our customers in the industry, as follows:

Gentlemen:

We are happy to announce that our strike of 100 days duration has been settled, and we will resume operations immediately.

While we have a heavy backlog of orders, we are in a position to give prompt service on new buckets and can, of course, ship any parts necessary immediately out of stock.

We wish to thank all of our friends and customers for their patience and forbearance during these trying months of uncertainty and labor turmoil, and to assure you that we will do our very best to give you the prompt and efficient service that we have tried to maintain for the past thirty years.

Yours very truly,
DRAKE-WILLIAMS-MOUNT COMPANY
Arthur W. Williams, V. President

These years were hard on Arthur, and not being able to get along well with his brother brought on a lot of stress. The strain of this probably was why he developed migraine headaches. These were very severe and would strike him often in the late afternoon or evening.

During the long strike, and also afterward, Arthur debated on an idea he had come up with of moving the company to another location. He had found a building in Fremont, and either had a lease or an option on it, and played with the moving plan. I suppose he knew that he could not walk away from the union contract, but thought that few of the men at the downtown plant would come along with the new location, and he might be able to possibly vote the union out after a time. Nothing came of it, and he must have let the option or the lease on that property run out.

The company made the paper again in 1949. The headline said, "Iowa Arrests 2 in Robbery." Two Minneapolis men were arrested and charged with robbery in a cabin camp in Red Oak, Iowa, and a third man was arrested in Minneapolis. They did an armed robbery near Tripoli, Iowa, taking $460 from a farm couple. The story continues to say that the three are wanted for a March 18 break-in at the office of Drake-Williams-Mount Co. When arrested, they were found with a typewriter and check writing machine. Of several payroll checks taken, about twenty were forged and cashed. They also had a .22 rifle in the cabin and two .38 caliber revolvers in their car.

The situation was not improving with Walt, however. His friends, his wife, and J.L. "Shorty" Snowden, our company auditor and desk man, advised him to go ahead and split up with his brother, but he never could quite make that decision. Somehow, he never seemed to be able to make the break.

Ernst Mount had worked right up to the end of his life. He died in 1934, and as noted, wrote his own obituary on his way to the hospital, from where he never returned and where he died.

Upon his death, Walter H. Williams was made president of the company and Arthur W. Williams, vice-president. My aunt and their sister, Georgia, had been active in the company. She acted as bookkeeper and probably wore many other hats as well. Before Hugh died, she acted as chauffeur for him and looked after his needs in many ways. I believe she was Hugh's favorite, son or not, and even though I did not know her for a long time, and I was quite small, I know she had a wonderful personality and great sense of humor. So I really do think she would have been Hugh's favorite child.

My cousin, Georgia Lee, believed our Aunt Georgia played a major part in the company in the years that she was there. Georgia Lee believes that our aunt held the place together, since there was so much animosity between the brothers. Unfortunately, Georgia died prematurely, in 1941 at the age of forty-six.

Both of the girls, Georgia and her sister Alice, had very good heads for business. Alice was a strong-willed lady, and she ran all three of the Benson-Williams Drug Stores in her day. That she ran those stores is for sure. Her word was never disputed in any way, and her orders were orders. She had all men (pharmacists, all) working under her, and all capable people, but Alice was indeed the boss.

The drugstores were bought by Hughie, who apparently felt that since the boys, Walter and Arthur, were succeeding to the business, that she should have a business for herself, and for her husband, Bert Benson. There was one at 42nd and Leavenworth Streets, one at 32nd Avenue and Arbor Street, and the main store at 35th and Leavenworth streets. These stores did very well in their day and always made a profit.

Other office personnel were J.L. "Shorty" Snowden and V.H. (Virginia) Huntington. Outside of the salesman, O'Donnel, who was there briefly, I do not know of any others, so it was a very trim, lean office to say the least.

Shorty was a bookkeeper, and an all-around sales clerk; and Virginia was a stenographer and secretary. She could actually take a little shorthand and, of course, type. In this, she was really kept busy, because at that time, all of our invoices had to be individually typed, and on top of this, monthly statements, also typed separately, were sent out. It was well into the late 1950s before we dispensed with the statements. She also did the complete payroll and had many other clerical duties. In those laid back days, many of our orders were just sent in by mail, and there was not a lot of phone business. Different times, and hard to imagine how much slower things were then.

In those years, if the company did manage to make a little profit, the brothers Williams would sit down in December and decide on an amount to set aside for a stock dividend. Of course, this meant that the money that they allowed themselves had already been taxed, once with the corporate income tax, and then they had to pay the personal income tax again on the final amount. It would have been very simple for them to agree on a special bonus, and those sums would be deductible, just as wages, but they had no professional financial assistance available to them and so probably never realized this.

Just about everything we know of in this life has a cycle—a youth and a time of growth, a maturity, and finally an end. Such has it been with the history of the Omaha dragline bucket.

The *World-Herald* article in the mid 1940s mentioned that when George Drake died in 1905 (wrong date, it should have said 1902) that the money the Williamses paid the George Drake estate figured largely in the construction of the Drake Court Apartments. I believe that the writer of the article got this information from Walter, because I do not think Art had that piece of information.

The article also related that all of the ownership of stock is in the Williams family, except some which is retained by the widow of Ernst Mount. So, some ten plus years had passed, and stock still remained in the widow's name.

In fact, I have Art's file on the estate of E.O. Mount, and Art and Walt were making the last payments on the shares to the Omaha National Bank, where the trust for the benefit of Mrs. Mount was apparently held. The last payments were not being made until 1954, twenty years following his death.

Mrs. Mount was Millie Grosbeck who was born in Chicago in 1884, therefore only about fifty years of age at Mount's death. She likely was still living when the final payment to the estate through Omaha National was made in 1954. She would be seventy at that time. Mount's mother and father were born in Germany and, according to *Who's Who in Omaha*, were William and Sophie. His mother, Sophie, came to Omaha, dying here in 1926. Mount lived until his death at

1335 S. 30th Ave., very near Hanscom Park, and also near the Gerald Ford birthsite on 32nd Avenue. It certainly was within walking distance of our 23rd and Hickory office.

When I worked in the summers while I was still a student, the company was making quite a few pressure vessels. We were approved to build ASME Code tanks and vessels, and this code was fairly stringent. This stands for American Society of Mechanical Engineers, and these vessels had to undergo two inspections as they were manufactured: One when the shell was rolled and welded up and the heads were there, ready to be welded in place, and two when the job was finished, and the tank was filled with water and put under pressure. This insured that there were no pinhole leaks, and that all was first-class. We made many of these for the York Corporation, and several others, and we seemed to do quite well with them.

I worked with one of our code welders, Yano Caniglia. He was an excellent welder, and over the noon hours, he helped me learn to weld. I never became proficient, but I could strike an arc. However, I never did any production welding. I'm sure the company made more money on these jobs than it did on building buckets.

Many years later, we lost out on this business when we failed to stay current with welding technology and methods. Other firms standardized on sizes and began building tanks in quantity. We were trying to compete doing ones and twos at most. It was for the best, however, since we were much better off when we reinvented ourselves and began to learn the ropes of the structural game.

Plus, we were miles ahead when it no longer became necessary to peen the welds in the vessel's welded joints. This took forever and was ear splitting. No wonder everyone who worked in that shop was hard of hearing!

Striving for New Business – the Elegrader

Another attempt at acquiring a proprietary product came in 1948 or 1949. Walter was still president, but this time, he did not object to the effort and the risk. The man's name was Verne Reisser, and he had been an officer with the American Road Equipment Co., manufacturers of various kinds of construction equipment in north Omaha. Verne was an engineer and the inventor of an item he called the Elegrader. Apparently, American management did not want to pursue the manufacture of this new item in their shops. They perhaps had all they could handle in making their "high-lift" equipment.

Verne's baby was an attachment for a road grader, or patrol. This was a device that had a disc, or plow share, which was located just on the edge of the blade or plow that all graders were equipped with. Off of the disc was a belt conveyor on rollers that was hinged so that it could be raised or lowered to any desired height. This enabled the operator to spill the dirt or material, such as at the edge of county or dirt roads, up and onto the conveyor belt, and from there it could be wind-rowed along the line of travel, or it could be loaded onto trucks that could follow along at the same rate of speed.

It was actually a pretty clever invention and had some practical applications. Brochures were printed up, and solicitations were made through the same dealer network that the company had established for Omaha buckets over the years. Many new dealers were added, since it was an item for inclusion with motor graders. Quite a bit of interest was shown in the early days, and orders, while not overwhelming, were coming in.

We had a book of forty or so dealers that sold buckets (not nearly enough) for us, and we circularized these, along with all of the Adams grader dealers, with the following letter, written, I am sure, by Art:

Advertising piece for the Elegrader, about 1950.

To All Adams Dealers:

We are enclosing a NEW bulletin and price list on a NEW piece of equipment developed by a NEW company. The new ELEGRADER is an elevating grader attachment for the Adams Models 550 and 610 motor graders.

The personnel of this new concern, Williams & Reisser Co., consists of the executive management of Drake-Williams-Mount Company. It is a partnership consisting of Walter H. Williams, Arthur W. Williams and Vernon H. Reisser. The first two named have spent their entire adult lives with Drake-Williams-Mount Co. and represents over 65 years of combined experience in the equipment field. Vernon H. Reisser joined

Drake-Williams-Mount Co. in August of 1950 after fifteen years of successful association with the American Road Equipment Co. which he founded in 1937 and headed until joining our company. Thus you can see that while Williams & Reisser Co. is new in business, it is old in experience, and has also enjoyed the full cooperation and help of the J. D. Adams Manufacturing Co.

This new product, the ELEGRADER is a unique piece of equipment and stands alone in the field, having many exclusive features found in no other Machine and outlined in the bulletin enclosed. We wish to point out particularly the folding carrier, trough type belt and self-lubricated sealed for life bearings on the troughing rollers. All of these features and many more will be helpful in selling this equipment against existing equipment. It is a truly fine piece of well-engineered, quality-built equipment and a fit companion for Adams motor graders.

An intensive advertising campaign is in the making, and we are enclosing a news release which has been sent, together with pictures, to every major construction magazine in the country, both national and regional. A full-page ad will appear in the January issue of 'Construction Equipment' which will be distributed to all members of the A.E.D. attending the convention in Chicago in January, so we believe that with this publicity and advertising, it won't be long until the ELEGRADER will become a well known name in the equipment field.

> We are now in production of this machine, but because of delays in tooling and obtaining outside supplies, production for early spring and summer will be limited. We therefore suggest that you anticipate your needs and place your orders as soon as possible.
>
> If additional copies of this bulletin or any other information are wanted, address your inquiries to Williams & Reisser Co., P.O. Box 1126, Omaha, Nebraska.

The enclosed price list covered three basic models, all for installation in the field, and were priced from $4,850 to $5,250, and included all the belts, bearings, shafts and couplings, and related parts, and, with the shipping weights given, were selling around 70 to 80 cents per pound! Also, no doubt, discount from these prices for the dealer made the prices even cheaper.

There were problems, of course. Probably the most serious was that Verne had made a deal with Adams, maker of a grader similar to Caterpillar's model. Adams was never able to effectively compete with Cat, and so this was a huge disadvantage. If Caterpillar would have been sold on the idea and the product, it could have really been big. We would have been blown away with orders and would not have had the shop space to keep up with demand.

Another vexing problem was centered on Verne himself. He was an inveterate tinkerer, as an engineer. He made modification upon modification, never satisfied with the latest design. This made it difficult to respond to parts requests from the field, since so many changes were continually being made. The documentation was wanting, probably because there was not much backup in the office for Verne.

About this time, Yano Caniglia, of whom I wrote earlier, had a very serious accident while testing the machine. He caught one arm in one of the drive belts, and his arm was almost completely severed. His was a long period of recuperation, but finally he got nearly the full use of the arm back again.

Verne traveled extensively, and his travel expenses were large. Being away, again the lack of backup support was a weakness and a sore spot. Also, Verne had a wife that apparently was not very well, and he got many, many calls and demands from home upon his time. Friction began to develop, and progress was really slow, if even existent. It was an unfortunate situation, and no one could say what "might have been."

At any rate, the principals finally agreed to call it quits. Verne had made a deal with Paxton-Mitchell, which had an operation called Jebco up in Blair, Nebraska. When the break was made, Verne left with certain equipment that was peculiar to the Elegrader, as well as all the parts and materials that were in inventory.

I can only imagine the disappointment this entailed for Art, since it had looked like such a promising new avenue for business and a badly needed item for the growth of the business. Of course, as time went by, he realized that the bucket business was also winding down and that we were having a tough time competing with Hendrix in the south. As usual, he kept these matters internalized and never showed how these disappointments really affected him. These things, plus the negative relations with his brother, really took a toll on him.

Verne also had owned a few shares of D-W-M capital stock, and here is where the break was not cleanly made. He still owned some of the shares, perhaps all, since I never saw any records of anyone buying shares back from Verne. Larry McMahon, at that time a part owner of the firm and vice president of our company, had just been killed in an auto accident on West Dodge Street. This was what prompted Verne to show up much later, on the very day of Larry McMahon's funeral, to see if we wouldn't now buy them back. (Larry's story is coming up.) Of course, I thought he could have at least waited a week or two, out of just a minimum of respect for the deceased, and for us in our loss.

Around this time, June of 1950, North Korea poured across the 38th parallel which was the common border with South Korea, and full-scale war broke out. The action brought a temporary increase in orders for our buckets, which meant parts orders as well. We had no other business in connection with this war effort, and the up tick in business was short lived.

When Walter had his heart attack, Art was really left alone in top management, and he began to cast about to try to find a solution. Certainly, he had to be still shell-shocked over efforts to find quality individuals who could shore up the business and really make contributions to its future. The company was wallowing, and business dwindling. He had to know that there was no real future in the boiler repair and tank business and that the era of steam was winding down. It is hard to know how much that Art ascribed to these factors, or how much he laid off to the generally "flat" economy.

Larry McMahon and Gate City Steel

It was Shorty Snowden who brought Larry McMahon to Art's attention, although the circumstances of this are unknown. It is likely that Larry found out that Shorty was employed part-time at the company. At that time, he worked also at Cris Rexall Drugstore on 50th and Dodge Streets.

Larry had quite a background in the industry. Years before, he had worked as a salesman for Ceco Steel, selling their lines of lightweight galvanized steel sheet, rain goods, and miscellaneous products, and, as such, he traveled the whole state of Nebraska, mostly by train, from town to town.

In the 1930s, he went to work for Henry Neff, who had formed Gate City Steel Co. This was a very fortunate move for Larry, getting in on the ground floor in a company with a great future, run by Henry and others, and that literally exploded with growth from the start. Larry became a vice-president in short order, heading up purchasing. He was doing the buying during the war years and found some truly innovative and creative ways to obtain steel, which of course was in critically short supply.

Several things sort of kept Gate City Steel in a class by itself. One was the excellent management team in place there, and people like Rolla Smith. Smith made it a habit to walk through the shops often. He made it his business to know every employee by his first name and to know something about his wife and family. Henry Neff himself also knew all his employees firsthand.

Second was the economy of Omaha at that time, which was buoyant and upbeat for building, both residential and commercial/industrial. Third was

Lawrence D. (Larry) McMahon came from Gate City Steel in 1956. Killed in auto crash, 1965.

the legendary profit-sharing plan that they maintained, which at that time was just about unheard of among businesses in our area. The plan was quite simple in an era with little government regulation or interference. In fact, I am pretty sure it was not even a written plan, and top management apparently just divided up whatever they decided should go into the pool at the end of the year, as they saw fit, trying to be as equitable as possible. This took place appropriately enough around Christmas time at the annual party, which was held for all the employees.

Some of these affairs got pretty raucous and tended to get out of hand. Eventually, the amount of drinking had to be curtailed, and mostly things were kept under control. Later, when Husky Oil of Canada bought out the company, everything changed. Gone were the annual celebrations, and gone was the profit-sharing plan itself. This marked the beginning of the end for Gate City, as numerous changes in management personnel were made, and the esprit that had been the trademark of Gate City people was soon gone. Everything changed.

Larry was now out of a job, either by his own choice, or at the convenience of the new owners, Husky Oil Co. and Glenn Nielsen. Larry organized a small company of his own, Nu-Steel, Inc., and had ideas of becoming a steel warehouse (distributor). He had an inventory of steel products, consisting mainly of house beams, pipe for pipe columns, and quite a lot of UM plate, which he could cut up for base and cap plates, and lintels. This steel was stored outside and, as such, was pretty rusty.

A few meetings were arranged between Art and Larry, and Larry proposed to "buy out" Walt's interests, using his inventory of steel, plus some cash and promissory notes to Walter. This made the decision for Art even a little harder to get around, because he felt he had to bargain for the interests of both Walt and himself.

Walt Retires Officially

Art wrestled with this decision in the final months of 1955 and finally took what really was his only option: to team up with Larry McMahon. The situation was much the same for Walter, as he was trying to recuperate from his heart attacks, and he had no other options, or buyers, for his interests. Here was what he had released to the newspaper:

> Arthur W. Williams today announced the retirement of Walter H. Williams, president of Drake-Williams-Mount Company, Omaha. Arthur W. Williams will become the new president and L.D. McMahon vice-president. Mr. McMahon has purchased the holdings of the retiring Walter Williams.
>
> Mr. McMahon is owner of Nu-Steel Company, Omaha, whose steel inventory will be consolidated with that of Drake-Williams-Mount.
>
> With 32 years of experience in the sales and warehousing of steel, Mr. McMahon recently resigned as executive vice president and general manager of Gate City Steel to enter business for himself.
>
> "The merger is one step in our plan to expand our steel warehousing," said the new company president, Arthur Williams. "We will stock a full line of hot rolled carbon bars and shapes, structural angles, beams, channels, reinforcing steel and welded mesh, steel

sheets and plates, black and galvanized pipe, steel boiler and condenser tubes." He went on to point out that Drake-Williams-Mount will continue the fabrication of all kinds of pressure vessels, components for high pressure refrigeration equipment, and will contract heavy steel plate and heavy sheet metal work. The company will also continue the manufacture of the 'Omaha' Bucket … the original lightweight dragline bucket now known and used in nearly every country in the world.

Retiring president Walter H. Williams has served with the company for 50 years. He was instrumental in the creation and development of the world famous "Omaha Bucket." He started working for the company at the wage of 5-cents per hour. He is resigning for reasons of ill health. New president Arthur Williams, who has been with the company for the past 30 years, expressed his assurance that he and Mr. McMahon would continue the tradition established by his brother of bending over backwards to serve fully the public and territory which provided the opportunity for growth to its present status.

Larry, Home Building Business, New Directions

This new era under the leadership of Art and Larry got off to a good start. Larry had some connections in the area among the home builders, and the largest of these was the Maenner Co. Larry also had a couple or three employees that he brought with him, but the best was one that I'm sure Larry didn't know the true value of, and that was Duane Shupe, who was to give us so many years of valuable and loyal service.

Somewhere, Larry had picked up a "salesman" who he turned loose on the home building industry. His name was Tex, and he was a real beaut! He sold house beams and columns mainly and didn't bother to get anybody's approval on prices. He would meet any and all prices on the word of the contractor, so when receivables began to back up, we found out that many of the builders were waiting for credits, because they were not getting "Tex's prices." Of course, they made you chase them down, and the price disparity was a good excuse for cruising along on someone else's money for as long as possible. Tex didn't last long, and he took off for parts unknown, probably thinking we owed him money.

But Tex's accounts were the little guys, and we enjoyed a lot of good business from T.H. Maenner in 1955 and 1956, but soon after this, they quit building tract homes. Almost immediately, Standard Iron Works began making inroads and soon had almost all of the home builder business. They were extremely low overhead and worked out of not much more than a shed, with their inventory spread out in a small yard out north. They delivered beams to the job site in a boom truck and, for no extra charge, would set the beams on the foundations for the contractor.

Larry had some other accounts among his friends, such as Ryan Construction (who did mainly industrial jobs, such as grain storage facilities) and Hollinsworth Construction, and Cudahy Packing Co., and, to some extent, Mainelli Construction Co. We had no support for furnishing bids and proposals for general work, so we were at a disadvantage at first.

Later, Larry's good friend, Simon Korshoj of Blair, became a good customer. With all of his faults, Larry was just what we needed at the right time in our history. He drank—true—and because he did, he was on the defensive much of the time. But Larry had that knowledge of the outside business world, including the steel industry, that we needed.

We were all Williamses and much too limited to the narrow facets of our own business and just very unsophisticated when it came to purchasing materials and dealing with customers. Larry needed partners within our company that he could work with, but we were not ready or experienced to fill that role. Art was not that person either, since he was not customer oriented and had been an administrator for too long. Later, when Mike and I found our "groove," things were beginning to come together and bear fruit. But then Larry went out and got himself killed.

We also lacked many other things in the form of cranes, equipment, tools, and rolling stock, and of course shop space. We had the old original shop that was moved from 19th and Pierce with its dirt floor and no heat. In periods of pouring rain, we accumulated mud on the floors that covered the plate steel that was stored on the floor, and we just had to wait for things to dry out.

This was the state of things when I joined the company after my stint in the Army, in October of 1956. I really don't think I had much idea of the magnitude of the work and the obstacles that lay ahead at that time.

We had acquired a union, as we have seen before, and there was not a positive relationship between the men and the office. A lot of mistrust on both sides, more by a lack of communication than anything else.

On the positive side, the company did have a good, if very limited, credit rating, and our capital strength was very small. We had a long and satisfactory history with the First National Bank, and Art had a personal good record with the Omaha National, which at that time was clearly the largest bank in town. The company, through very conservative (reactionary?) management, had survived the Great Depression, one in which Art and Walt had witnessed a lot of other companies wash up on the beach. Relationships with suppliers were good, but the company had no steel mill connections, except the very limited one with Inland Steel Co.

In addition, the U.S. had fought a second World War, plus the conflict in Korea, and these two events may have saved the company with the added business that this entailed, as they also saved the country itself. An eventful decade, the 1950s: Ike, discovery of penicillin, "Peanuts," McDonalds and KFC, Playboy, McCarthy hearings, Disneyland, Dr. Seuss, computers, and thankfully, goodbye Joseph Stalin.

Arthur assumed the office of president, and I was there when he said to Virginia, Shorty, and me: "I just knew if I hung around here long enough, I'd be president!"

Meanwhile, my thoughts were directed almost entirely at sales, as green as I was, and figuring out how to get new business. The company was entering a new era, and the new year was 1956—and we had survived!

"Under New Management"

I tried to learn all about the bucket business, and to know the nomenclature of the parts, plus the associated part number. Even though I had helped in the shop in the assembly, I struggled with the various part names and numbers. I had a hard time getting enthused about this line of business, I think because there was no exposure to the ultimate customer, and the product was used in a remote area by unseen and unknown persons, and what was there to talk about?

Far more important to the customer/contractor had to be the crane that manipulated the bucket on the end of the boom. The dealer, our customer, was far more interested in selling the crane—the big ticket item—than the secondary item, or the bucket. He made fifty times more money selling the crane and usually could not care less whose bucket went along with the sale.

I was much more attracted to the sale of steel and various fabrications. The only problem was, I didn't have a clear idea of where all the steel was being used and who the potential customers were! I went through the yellow pages and other publications, and made endless lists. I got a thick blue book with removable pages and made pages for everyone that I thought could possibly use steel in "warehouse" form. Many of the entries were laughable, but I checked them all out. I arranged the pages in accordance with where they were located—Southeast, Northeast, Southwest, and Northwest.

I made countless fruitless calls and ran up all kinds of blind alleys. I wasn't selling anything, but still, it was a learning experience. I learned lots of things not to do, and lots of places to not waste time on. We could not sell the sheet metal shops, because we did not have galvanized sheets. We did not have cold rolled to any extent, not tool steel, so we could not do much with machine shops.

I was not out on calls full-time, which was probably a good thing, since I was not being effective and had so much to learn. I wanted some kind of a new image for our company and wanted us to shorten the name to simply "Drake-Williams Steel"—to drop the "Mount" since there was no Mount and had not been for years.

I worked on a logo and played with a number of ideas. I had some experience with art, lettering, and graphics and soon came up with the logo that we used for years on our letterheads and advertising: "House of Steel." This was just okay, but some of us complained that it suggested we were suppliers to the home building industry or related activities. Nevertheless, we used it for years, until I designed the new one, which we still use today.

We were fast evolving into the Third Generation of Williams influence and direction, but that is for the next chapter of this story.

THREE

THIRD GENERATION: BOOM TIMES AND A NEW DIRECTION

IT WAS 1954, AROUND CHRISTMAS TIME, AND MARY JEAN AND I were still becoming accustomed to being proud new parents of our son, David Wade, born November 24. The U.S. did not want me to enjoy this special time apparently, and I had a date with the Army, and my "invitation" stated that I should present myself at Ft. Leonard Wood, Missouri, on January 10, 1955. I was not really looking forward to two years out of our young married life, but we were not alone in this regard. Most of our friends ended up serving in the armed forces as well. I felt fortunate that, at the least, we were not engaged in a shooting war at the time.

We were having something of a final "fling" and were occupying one of the bridal suites at the old Blackstone Hotel, compliments of Mary Jean's overly generous father, Marshall, as a sort of second honeymoon. Or perhaps it was to cushion the shock of my imminent indoctrination into basic training, U.S. Army style.

My sister, Janet, and her husband, Delno, were in town, so we invited them up to see our really fancy digs and to have a little quiet, private party. We sat around and talked about a lot of things, and eventually Janet asked me what my plans were after I got through with my military obligation.

Even though I had yet to really work full-time at the "shop" (as we referred to it at that time), I told her that my ambition was to come back and work for the company and to build something there—that I felt that I would like to come back and save the company. I knew it had been a lot larger and more prosperous, but it was now sinking slowly out of sight.

Janet tried to discourage me immediately. She talked about the fact that Dad and Walt (our uncle Walter Williams) never got along together and were constantly bickering, which made any kind of progress for the company nearly impossible. She said it held no future for me. Nevertheless, I told her that was what I would do, and I really felt "called" to try to make a difference in the company's future. Maybe it was because I carried the name of one of the founders—Hugh. I guess she didn't think I could really have any big effect there, and I would be wasting my time.

Thinking about that now, I really wonder just what my exact thoughts were. I knew little about what we really did at Drake-Williams-Mount Co., even though I had worked there two summers. It must have had something to do with the arrogance of youth, I guess. Janet's reaction, however, was that the company was a deadend and going nowhere. She said it was dirty, old, and a pit, or words to that effect. She urged me to reconsider and said I would be throwing my whole life away!

Anyway, I was determined that I would give it my best shot. Her knowledge of the place was even sketchier than mine, and about all she remembered was being in the office where she would come from high school and be waiting for Dad for the long ride home. We lived at 105th Street, a long way out of the city limits at that time She might have been right about a lot of things, except for events that were soon to come about and have big effects on the future of the company.

Walt Williams, as has been stated, was Art's older brother, and it is true, they rarely were in agreement on anything. As reported earlier, Walter was beyond conservative—reactionary, really. He lived frugally and felt that the frugality should extend to the business as well. He came up in the era of steam power, was probably scared to his innards by the business failures he witnessed during the Great Depression, and he also was not well schooled. He was largely self-taught as a mechanical engineer, receiving his engineering certification from the State of Nebraska under the "grandfather" clause.

Although having such a severe personality, Walt had lots of friends and cronies in the business, and in the packing houses, and loved to keep track of the day's "hog kill" and the number of cattle and sheep processed. He loved boilers and breechings, smokestacks, and anything made of sheet metal. He loved repairing things, as if he loved it because it was "preferable to the idea of investing in anything new." He would say "bricks and mortar," and he always predicted the imminent and total collapse of the economy. Gloom and doom were his invariable reactions. Well, we were all conservative in our politics, but Walt was far beyond that. He was negative—even beyond that to, as I say—actually reactionary.

Naturally, since I got most of my information from Art, I always got his side of every dispute, every argument with Walter. So, Walt was always the curmudgeon, always the stick-in-the-mud, always the obstructer of progress. Walt, when he stopped to see Lizzie (their mother), would limp badly and wheeze and puff and tell her how exhausted he was, and how he was really the only thing keeping the "shop" alive, and intimate that Art did very little to hold up his end of their "partnership."

It got so the families barely spoke—Walt and Ruth, on the one side, Art and Janet the other. Walt and Ruth were alienated from other family members as well, including Alice, a sister, and her husband, Bert, plus their only daughter, Georgia Lee—despite the fact that they lived just a few houses away on Parker Street.

Walter Williams Era Ends

In 1955, Walt had a pretty severe heart attack and was forced to retire as president and from any active kind of a role with the company. Walt's son, Walt, Jr. (Bud), worked in the shop and had been ever since his separation from the military in about 1945. In over ten years, Walt had never seen fit to try to bring Bud up to the office and to begin to teach him something about the management of the business.

So, after Walt's illness, the office force was left very thin. In fact, you might say at the bare bones level. In addition to my father, Art, there was Virginia Huntington, and J.L. "Shorty" Snowden, a bookkeeper. And Shorty was only able to give them part-time hours, due to other commitments. This also speaks volumes of the level of business and of confidence at that point in time for Drake-Williams-Mount Co.

Bud was a very qualified fitter-welder in the shop and was certainly needed there, but Art knew that he should come up in the office at this time and, at long last, learn the administrative and business side of operations. Besides, they were so undermanned in the office. Art called him in, and he agreed to the change in responsibilities. Dutifully, Bud came in, white shirt and tie instead of the accustomed blue cotton work shirt and apron of the shop. He was expected to man the phone, respond to customer inquiries, and estimate jobs, such as tanks and pressure vessels—things that he had been working on and building for the better part of ten years.

Walter later returned on a part-time basis, but during this period was apparently not able to be there during a time that was proving to be very critical for Bud. The result of this was that Bud received no training at all in the skills needed to man the phones, do sales or estimating, or any other work of an administrative nature. How much time Art spent with him is anybody's guess, but I know that my father was not the most patient of teachers.

I was in the service from January 1955 until I received an early release in October 1956. I was not there, so I do not know what the exact situation was with Bud, but Art told me that Bud was really uncomfortable in his new position, and that finally, after several weeks, he came into Art's office and told him that he could not continue, that he was being physically sick in the mornings, and that he wanted to resign and go elsewhere to work. It was a very unfortunate turn of events for Bud, and I can only guess that he thought the business was going down the tubes and that he did not want to have any part of it. He had no stomach for the battle that he thought he could see coming.

Bud was a very nice guy—very quiet and shy—introverted, certainly. His mother, Ruth, was a very emotional and high-strung woman, and not at all similar in personality to Walter. She wrote poetry, and when their daughter, Elaine, who was younger than Bud, died, she wrote a whole book of poems around her brief life. Bud's makeup resembled his mother's much more so than Walt's, and Walter took advantage of this weakness in his personality and made decisions for Bud, even long after he was married and living apart from his parents. Bud never seemed to question his father, and his word was always "the law." Walt knew this, but he seemed to enjoy the "hold" that he had over poor Bud.

Art felt that brother Walt, being domineering and overbearing, weighed heavily on poor Bud's personality. I remember them talking about the fact that Bud, at maybe age thirty-three, could not make the decision on what car he would buy, but rather Walter would decide that for him. And, of course, the fact was that Walt kept him in the shop all those years, not taking him in hand and teaching him some of the ropes.

Bud left the company, however, and took a job with T.S. McShane & Co., downtown, who were machinery dealers. They needed a welder to do various add-ons and modifications to equipment that they had sold. Bud worked happily here for quite a while and then later took a position with the Donaldson Corporation. This was selling industrial accounts parts and various standard items for industrial equipment. He carried around a catalog at least 5 inches thick, and a huge briefcase, indicative of the thousands of items they carried.

Bud perhaps threw away a chance at a percentage of ownership in the company; however, no one really knows whether Walter would have given him the chance or not. Did Walter offer Bud any chance at a percentage of ownership? Did Bud turn down an offer to be an owner? We will never know the answers to those questions, but I know that Mike Williams and I were not unhappy to have Bud out of the picture, and certainly did not want him as a partner.

All of this resulted in perhaps the lowest point in the long history of the company, in that there was so little in the form of management and leadership of the company left. The turn of events left Art still very short in the office, and what he really wanted was a partner of sorts, someone to shoulder a share of the responsibilities, as well as the risks.

Up to this time, I had worked in the shop a couple of summers, on my vacation from high school. At that time, we had about twenty shop employees. Our facilities consisted of a quarter-inch Cincinnati plate shear, a quarter-inch Dreiss and Krump apron brake, a couple of old single end plate punches, the 12 foot Pels plate roll, an Austin-Western 5-ton yard crane, and no overhead cranes at all. We had some old machine lathes, a 12 foot pyramid-type plate rolls, a shorter (homemade) plate roll, and very little else.

We operated out of two shops. The original (this was the one moved from 19th and Pierce in about 1901) was about 50 feet by 200 feet, with a dirt floor. The floor sloped from north to south, was unheated, and missing many windows. The West shop was of brick, and was newer, and was about 45 feet by 180 feet, plus a wooden lean-to shed on the east side where we stored boiler tubes. We owned one International flat-bed straight truck, which was at least fifteen years old. Our driver was old Oscar Boonstra, who was steady, but never moved very fast.

The office was about 30 feet square, two rooms, and one toilet room (unisex before its time). We had no connections with the steel mills, with the exception of one—Inland Steel—from which we purchased a few steel plates for our pressure vessel work. I remember their rep, a real old-timer and quite a gentleman by the name of Henry Darby. He lived and worked out of Kansas City. Whenever he came by, which was about once a year, my Dad treated him with great deference and respect. He was impeccably dressed and, with his pure white hair, looked like a banker when he called.

I am sure that Inland probably kept a pretty close eye on our account. We rarely borrowed money from our bank, the First National of Omaha. But we did have a clean credit history. We were careful, we did not spend money we did not have, and we paid our bills on time. Every December, if we had a "decent" year, Art and Walt would declare a small dividend, which meant the funds got taxed yet again before they got hold of them. Many years, though, the dividend was passed. Needless to say, the company was never audited by the IRS.

The Larry McMahon Years

In late 1955, I received orders from the Army to go to Japan. Home on leave, I was down at the office when I first met Larry McMahon. Larry had a career that had included selling on the road for Ceco Steel, and then, at Gate City Steel, he had been vice-president of purchasing. He had been there for the "glory days" of Gate City, with Henry Neff, the founder, and Ivan Clough, Rolla Smith, Leonidas Warren, and many of the other key people.

Those salad days were done at Gate City, sometime after it was purchased by Husky Oil of Canada. Husky used Gate City for a cash cow, to feed their growing needs for oil exploration funds. All of the profit-sharing plans, which had been such a rousing success, and a very key motivator, were cancelled.

The story was that Larry and several others had been let go in the shuffle. Somehow, through Shorty Snowden, Larry was put in touch with Art, and they were in discussions relating to the possibility of Larry buying Walter's stock and certain company Treasury Stock. Larry did not have much cash, but he had an inventory of steel from Nu-Steel, a company that he had founded and had operated for several years at 13th and Grace Streets. Larry would make a downpayment to Walt, and Walt would carry back the balance on a note for about five years. That was the deal, and that is essentially how it turned out, eventually. It is interesting that Walt had apparently given up on the possibility of Bud ever getting back in the picture, with some semblance of ownership.

I did not have any real experience, but meeting Larry for the first time did not impress me too much. He did not seem totally for "real," but he certainly had great experience that we badly needed. Art was really on tenterhooks on this one. He suffered some more of his migraines that afflicted him periodically. He really agonized over the decision. There were the stories about Larry's drinking. Well, we all liked our booze, right? Was he honest enough? Could he be trusted? He seems a likable enough Irishman—but then, he was all Irish!

I was in Japan when I heard that the deal had been closed. Soon, I got a couple of letters from my Dad (who had never written me in twenty-five prior years) talking about the business upturn and the good job McMahon was doing. I was really anxious, more than ever, to get home and to get started with this company.

I heard about a program for early release from the service for the purpose of going to school. I had already graduated from the University of Omaha, but the Army did not know this, and I applied anyway. We had a son who was going on two years old, and we found out we had another one on the way, due in February. I can get out of the service a whole three months early! Well, why not? I certainly was never going to make the Army my career.

After a seeming eternity of red tape, waiting, more forms, more waiting, I was finally released at Fort Chaffee, Arkansas. After the interminable delays, the final act was incredibly brief. We were called out into a formation of a few mostly antsy, unmilitary types, with an officer reading the orders of separation, and, bingo, it was all over. It was wildly anticlimactic. I wanted to celebrate, to come out with a huge cheer, but there was no one there that I knew to shout at. It was like a huge weight off my shoulders, and I could hardly believe I was finally free. I was a civilian, pretty much broke, but absolutely free, and then I high-tailed it for Omaha.

Aerial view of Drake-Williams-Mount Co. about 1950.

A New Broom Sweeps Clean

I remember the first day I walked in the office. Seven-thirty in the morning, no one there but Shorty. I talked to him for a few minutes, and then, since there were few other places to sit, sat down at Art's desk in the only other room off the main office. Later, Shorty told all the others that I apparently had decided to start at the top, right at the president's desk.

Mary Jean and I found an apartment on the second floor of a house on 54th and Mason Streets, and not long after that, on the first of February, John was born. We were suddenly a family of four. It wasn't to be too long before we were straining at the bit to find a small home to accommodate our needs. What a blessing to now have two fine boys.

Arthur Williams had some fine skills and talents, but teaching was not one of them. He did not have an overabundance of patience, and he was inclined to think that if there was something that he knew, then surely all those around him must know it too, through some kind of mental telepathy or osmosis perhaps. In fact, he could get downright irritable when anyone asked him some of these very basic things—things he figured you should damn well know.

The year 1956, to set the scene, was when Elvis Presley burst from nowhere. Coming home from Japan and the service, we thought the country had gone crazy and tried to figure out what anybody could possibly see in this "shimmying" so and so. Khrushchev repudiated Stalin and was in firm control in Russia, and he ruthlessly put down revolts in Poland and Hungary. We lost comedian Fred Allen; Connie Mack, baseball manager; Babe Dedrikson Zaharias, the great golfer; and H.L. Mencken, the great writer/editor.

George Gobel was laying them in the aisles, and Jimmy Durante was star of TV's *Texaco Theater*, where Milton Berle had once reigned. Patti Page warbled, and Harry Belafonte had people tapping feet to his brand of calypso music.

I wasn't all that aware of the popular scene, as I spent the first month or two learning strictly about clerical procedures and things as mundane as operation of simple business machines. My teacher was Virginia Huntington, our "girl Friday," of course. I spent my spare time trying to lay out a plan for outside selling and developing a list of prospects from everything I could get my hands on. I learned how to spread and post labor hours to each job cost sheet. No automation, all just done by hand.

Then I learned how to do the payroll, making it balance by hand, doing all the withholding and other deductions. It was a good thing for me that we had to do only about thirty people in those days. I got a very good firsthand appreciation of the gross dollars and the net dollars—in other words, the wages that "Uncle" was holding, even then, a real eye-opener. No spread sheets at that time, and, of course, what a time saver that would have been.

We had a couple of old mechanical calculator/adding machines. The electronics were yet to come, as were the first ever copying machines. Then, all the checks had to be typed out by hand on an old Underwood that had caps only. Doing that payroll took at least four hours to complete.

Even though Larry McMahon was with us in the capacity of vice-president, he was not around much in the office. He spent his time, ostensibly, chasing business. Much of the time, we found out that he was chasing it in saloons and bars. It was as if he avoided the office and saw himself as the outside sales force. But we had no backup in sales or estimating, and Larry did not know how to provide it. He was feeling his way, and the rules for what latitude he had under Art were probably not really clear to him.

For Art's part, I guess he thought that there was no reason that Larry should not just go out and rack up the orders, and we could grow with no growing pains, no sweat. Things just would not really work that way. If Art ever laid out a comprehensive plan for how all of the individual parts should function, and how we would then grow as a company, I was not aware of it. It seems "seat of the pants" was the order of the day.

ASARCO

One of our best customers, far and away, was ASARCO. For years we had been doing business with them. Much of it was heavy plate

Heavy plate steel lead kettle, one of many fabricated for ASARCO (American Smelting & Refining Co).

fabrication, and it was good, profitable work. The "boys" down at ASARCO were friendly toward us, and once or twice a year came to expect a big luncheon engagement. Art would meet them at Cantoni's restaurant at 19th and Leavenworth.

There was Frank Prucka, chief engineer, and Ernie Abersfeller, chief draftsman. Also, Max Dolleck, purchasing manager, and Harry Ramsey, manager of the machine shop. Lots of drinks and big, evening sized steaks, and back to the office about 3 to 4 in the afternoon. The ASARCO guys really looked forward to these affairs, and certainly, we were favored with lots of orders. I doubt these guys were paid that well, and this was a "perk" that they figured they had coming.

Many years earlier, we had fabricated and erected a very large self-supporting smokestack at ASARCO's plant. It was perhaps 300 feet or so in height and was among the tallest anywhere. It was before welding was prevalent, and so was entirely riveted. Many years later, the company became concerned that it might be a hazard due to corrosion and age, so they cut down about 100 feet of it, and they found no deterioration or major corrosion of any kind to the stack.

In those days, we used to keep the office open on Saturdays. We usually would have a couple of shop people until noon, also, to fill emergency orders for customers. I used to draw the Saturday duty and would use the time to catch up on estimating or sketching items up for shop orders. A frequent visitor on Saturdays would be old Frank Prucka (not to be confused with Hiram Prucka of Prucka Transportation). Frank was a really bad drunk, and on Saturday mornings, he apparently had no one for a drinking partner.

Art kept a couple of bottles of booze in a cabinet in the middle office. Frank knew exactly where they were and would just come in and help himself. Then he would sit down across from me and tell me how we should be running our business. He would castigate Walter for keeping his son Bud out there in the shop. Then he would instruct me about how we should get to growing the business and ask me if I knew anything about Gate City Steel, and about their history of growth over the last twenty years. I assured him that I did, and that we had plans of our own and plenty of aspirations for the future, as well. Frank hardly heard what I was saying, and I don't think he sensed the ambitions I had burning inside of me.

Poor Frank was on a fast track to nowhere. He was a pretty talented engineer and a good employee when he was sober. But his drinking got steadily worse. I remember one time his telling Art that, no matter how much he drank, he "couldn't get drunk anymore," and that he couldn't sleep at night and would get up in the middle of the night and get his bottle, which he kept in the freezer compartment, to let him get back to sleep. Soon, I was one of the mourners at his funeral. Once again, it seemed no one could help him, and the poor guy surely couldn't help himself.

Frank mentioned Gate City, and, yes, I certainly was aware of them. I looked at them with a mixture of admiration and plenty of envy. I remember the first time I was at their plant, probably to pick up some bolts or something. I walked over to one of the doors to the main east-west shop, which is about 770 feet long and 90 or 100 feet wide. I had never even seen an overhead crane in operation, and here was one, slowly rolling along the craneway with a huge girder sucked up under it. It was quite a sight to a young man just getting his feet wet. I never once dared to dream that some day we would own this plant and be building on even more capacity. It looked like all the money in the world to me and like something we could scarcely not dare to equal in size and scope.

I knew something of the story of Gate City's history, of the great incentives engendered by their profit-sharing plan. Rolla Smith was general manager, and the story was he could walk through the shop and call every man by his name, knew each of their wives, and even their families. Around Christmastime they would have a huge employee party, where a fair amount of celebrating would be going on, and checks would be distributed to all the employees. Total harmony, great morale—truly a joyful place to work.

The truth was, these things were now in the past, and all those plans and policies had been abruptly ended when Husky Oil bought the company. They put in their own people with their own set of ground rules. The euphoria was fast disappearing, but some of the good will had a momentum of its own for a time.

This was a lesson to me in itself. If you have a good employee, he deserves to keep part of the profits he is helping to create for the whole company. If he is doing his part, you better damn sure do your part. Don't ever take it away, but instead, put yourself in his place. Don't put limits on him. Give him a sense of ownership. Give him pride, praise, and give him a piece of the pie. Anything else, the employee figures he is in a deadend situation, and he soon figures there is no reason to knock himself out—after all, everybody is treated just about the same, no matter what kind of work they do, or how much effort they put out.

I'm sure Husky Oil could not admit the mistake they had made, and for the next ten years or more, they kept rotating new managers and executives in and out, but never could achieve anything like the results that they had counted on and that the old regime had always produced. The lesson for me at this result was one that I never forgot!

Scrapping for Business

I finally ventured out, dumb as dirt, and began cold calling on various people. I kept a fat book and jotted little notes about anything I could find of relevance that might help me next time to connect. I sold a few house beams to the home builders, but found out quickly that a company called Standard Iron Works sold the builders on a giveaway basis. I made a million false moves and called on everyone I could think of. I was given some past dues to collect and had success with this for the most part. But I could not seem to crack (what we called) the warehouse market. I called on the welding shops and sold Frank Tomasek down on 13th Street. I called on the trailer sales and repair shops. Western Transport would buy, but would not pay their bills. I finally had some success with Trailmobile, yet could not break into Fruehauf down the street.

Trailmobile needed special plate bending, and we could not always get it with our apron brake. So I would rush the material over to Harry

Olsen on North 24th Street, and they would make the bends, and we would make another trip and deliver it to Trailmobile. Talk about double handling!

At that time, we belonged to the trade association known as the American Steel Warehouse Association. Most all of the steel companies were members, even including the Baum Iron Co. from 13th and Harney. This was owned by the very eccentric Sloan Allen and his son, David. These characters were as if they had been transplanted from the Ivy League, as they wore their seersucker suits summer and winter. They lived somewhere in the vicinity of 40th and Dewey and could be seen of a morning walking all the way downtown to their offices—David following a few paces behind Sloan.

Even Omaha Steel Works was a member of ASWA. Then, along came a genius by the name of Bob Welch who changed the name to SSCI, the Steel Service Center Institute, and the outfit was never the same. He shot the dues through the roof and began to cater only to the huge warehouses (he insisted on calling them "service centers") and to the steel mills, because he could get lots of funds easily this way.

This all succeeded in breaking up the organization, and Omaha Steel and Baum immediately dropped out, as did we also, somewhat later.

I kept that vision of Gate City (and to some extent Omaha Steel and Paxton Vierling Steel) in my mind. I knew we were miles behind, and we had to scramble and scratch like hell to try to compete. I'm sure they were hardly aware of us at all.

Then I discovered Wilson Concrete Co., and all hell seemed to break loose. They were busy, busy, busy, and demanding, demanding, demanding. Omaha was beginning a building boom, and Wilson was right in the center of the storm. We caught hell at times, but we did tons of business, it seemed. We had so little equipment; we were always scrambling to meet their deadlines and demands. We had no backup in the way of estimating and detailing departments. We had no quality control, either, and a lot of mistakes came bouncing back on our single delivery truck.

Soon, I was spending about half my time out on the street, but as I developed more and more business, I had a rough time coping with the follow-up, the estimating, and administration. We had no one inside except Shorty, and he was swamped. We couldn't very well hire people, when we had no place to put them!

I was not getting out on sales calls enough, because of the necessity of follow-up on Wilson and other orders. We needed staff badly and were finding out how painful it was to be growing to meet the demand. We were at a huge disadvantage to our in-town competitors, because they had more inventory, more equipment, and more support staff, but we were able to beat them a fair amount of the time anyway. We did it with personal attention and service. We made mistakes, we got our butts kicked, but we did not give up. We just kept coming back, and we got a lot of business.

About that time, we decided on a huge office expansion. This amounted to the addition of one room onto the L-shaped office, filling it out to its eventual square shape. We had Del Chapin, who was Larry McMahon's brother-in-law, do the work. He was Korshoj Construction Co. in Omaha and ran the Omaha operation for Simon Korshoj of Blair. Not quite a 50 percent expansion; nevertheless, it afforded a private office for Art and did give us a little more "wiggle" room.

The chief feature of this office was the unique wall treatment. Del had a man who did plastering, and he was somewhat of an artist in that he worked color into the plaster, and then, before the plaster dried, and using a pointing tool and a long straight edge, made grooves that resembled courses of brickwork in all four walls. So we had a very fancy private office in the otherwise nondescript and ordinary old building. Then Art picked up a red leather couch that Happy Hollow Club was disposing of and bought a green leather chair, and had Janet, his wife and my mother, make some curtains for the two windows in the room.

The object of all of this is to underline how "poor" we were and how limited were our means to afford anything extra in the way of administrative expenses. Along about this time, we replaced the old hand-cranked gas pump that stood for years in front of the office. We bought a secondhand (electric) gas pump from someone in town. What an improvement!

Along with the "sweeping changes" to our offices, we had at that time a wooden porch on the north side, which was the main doorway. We decided that the porch was in pretty poor shape and that we could use a new one. No sense in making another out of wood for crying out loud, we would make a complete porch out of steel. Someone drew up a plan, and we put it in the shop one day.

George Spicka drew the job of welding the thing together. George was an old boilermaker, and he was our outside man if we had any boiler repairs to do. George's wife worked at Willsie Cap and Gown Co. downtown, so George would show up early most mornings, having dropped off his wife at her job. When he came in the office, Shorty would refer to him (since he had two paychecks in the family) as an "economic royalist"—a term associated mostly with the class warfare of earlier times.

But the story is about the new porch being readied, and it was on a Friday, which was payday for the men. At that time, Virginia would have the checks ready, and the men (there were only about thirty) would file through the office, coming up the back stairs, and then out the front door. The old porch had been torn out, and we had put a sign on the inside of the front door warning that there was no longer a porch outside the door. Now, George had spent the better part of the day putting the steel replacement porch together, but—you guessed it—when he got his check in the office, he proceeded straight to the front door and, without further ado, went ass over teakettle onto the concrete below. It was fortunate that he did not sustain serious injury, except to his ego.

When the first 3M Thermofax copying machines came out, it took quite a lot of lobbying and persuading on my part to get this big expense approved. It was a heat-sensitive system that used a special paper, almost like tissue paper, for the copy media. If a copy was exposed to strong light or any heat at all, the image was lost. Everyone smoked in those days, and cigarettes played a part in ruining a lot of copies. But we were thoroughly thrilled by this new invention. So many things that came to us as sketches could be copied this way and attached to our shop orders.

As other copy machines were developed, we constantly tried to come up with ways that we could streamline our paperwork. The big objective was to find a way to only need to write an order once. Then, the copying machine could produce audit copy, delivery tickets, perhaps bill of lading, and invoice, all at the same time. Progress, a veritable quantum leap!

Mike Williams to the Front

Mike Williams, my brother, came along in 1958. He worked for a year or more in the shop, learning all the operations and procedures that he could. He remembers working with old Tony McAndrews. Tony's cousin was Don McAndrews, who grew to be one of our best mechanics and lead men in the shop. They worked on the old Dreiss and Krump apron brake, and Tony always said Mike was a great helper and worker. Mike also remembers working with others, like Steve Juhas, who ran the big Pels 12-foot plate rolls.

Steve needed someone with plenty of muscle to knock down the initial front edge of the wide plates, so that they could get a start into the big top roll. This was done with "Monday," which was the name of the sledge hammer used to do the knocking down. It was called Monday because it weighed about 50 percent more than the other sledges, and the idea was, you had better use it on Monday, when you were fresh and strong! A few workouts like this, and you slept really well that night.

Steve Juhas was a small man, and brother-in-law to our long-time, but retired shop foreman, George Hoody. I remember how funny it seemed to me that Steve wore his old denim jacket all year round—no matter how cool or hot the weather. Mike and I would kid Steve and say, "Shteve, why don' you takea' you code off!" Steve would just shrug his shoulders and grin at you. "No," he'd say. I doubt he knew we were kidding him.

Mike Williams (brother to Hugh) was vice-president from 1965 until his retirement in 1993.

Mike spent that valuable time in the shop, where he needed the grounding. It was great when he was brought up into the office, and he soon learned estimating and was learning purchasing as well. Mike was really efficient with numbers and could knock out prodigious amounts of estimating. We would have races "weighing up" page after page of our material lists. Mike always won. We were scrupulous in our accounting for every piece of steel material but did not make estimates of manhours. We grouped items together and "pound priced" as we went along. Consequently, we had no idea of how long the various phases of the work would take. But we still needed help and backup for take-off work

Shorty Snowden was one of our "anchors" in the office. He was a real old-timer and always believed that the customer came first. You had to give him service with a capital S. He literally fought for his customers, and when he made them a delivery promise, he wanted backup and follow through by the shop.

One day, one of his promises was not materializing, and Shorty was plenty sore. We had a guy by the name of Dick Bowers who ran the Austin-Western yard crane for us, and he was also the straw boss of the warehouse crew, filling orders.

It was a very warm summer day, and Shorty was even hotter. He took off down the back stairs of our old office and headed straight for Dick. Bowers was plenty busy and was probably doing the best he could with what he had, but Shorty was in no mood for small talk. While Dick was sitting in his seat on the crane, Shorty stood below him, as tall as he could, but on the ground, and started chewing.

About halfway through this tirade, a big glob of grease decided to drop free of the crane boom, directly above old Shorty. It lit right on his shoulder, smack on his clean white shirt. Shorty didn't say another word, but just spun around and headed back for the office. For the rest of that day, Shorty, having wrung out his shirt with water and hung it to dry, sat at his desk, ducking behind it every time someone came to the front door. Of course, he was embarrassed as hell, but you could never accuse him of being timid about demanding service for his customers.

But Shorty was a really valuable part of our small office "team" at that time. He was pretty good on the phones with our customers and really was a company man, while at the same time trying to give prompt service to the customer. I'm afraid Mike and I (but especially Mike) kidded him to the verge of a heart attack in those years. In the beginning, he did all of our yearend accounting. Although really just a bookkeeper, he knew most of the rudiments of accounting, but the best thing that eventuated was that the First National Bank required a certified statement from a CPA, as part of the terms of the loan that we secured for our first expansion. This put us on a firm footing for the future years.

I remember at that time, I was imagining a new shop and trying to visualize how it might fit on our property. To me, the logical place to locate a shop like this was back at the rear of our ground, where the longest part of the property was. We needed as much space as possible to my mind. In the back of my mind, I knew there was always the chance to acquire the property to our west. We would have all the ground we would ever need and could build an "enormous" facility, in stages if necessary.

I fiddled around with plan drawings, wanting to see just how much of a building could be constructed there. I made a larger sketch on tracing paper and put in pieces of equipment where I thought they should be and approximately in scale. I showed this to Mike, and I could see he was really excited about the idea as well.

We talked about what we would like to accomplish, but we knew we were stymied for the time being. I know that this was when Mike first became excited about growing and expanding the business. Right there and then, we had an unspoken partnership, and we were on the "same page" in our thinking. We wanted in the worst way to take the company out of the "dark ages" and to make some real headway. Our ideas were vague and sketchy. We only knew we wanted to "get there." We would bide our time. We were young, but time was on our side.

Howard Coe

At some point, along came Howard Coe. He had worked for a time for Paxton-Vierling Steel. This came about because Howard had actually introduced old Fred Owens to the Vierling people, and that was what later developed into the sale of that company to the Owens family. He was working for A.M. Castle, a Chicago warehouse steel supplier, traveling Nebraska strictly on commission, so he needed another principal, someone who was close and who could do a little fabrication. He developed a few accounts, but never set the world on fire, but soon he was working solely for us.

Howard was a salesman out of the old school, so he fit in well. He passed out cheap cigars to his customers, but drove right past many of the big buyers. If the customer gave him a hard time, or asked too many questions about price, he would just avoid him. He did not know how to price material, and he was not interested in learning. He knew even less about fabrication, but he was well liked and had success with quite a few outstate accounts.

Probably we never made money on these outstate efforts, but the whole idea was outmoded. The competition was running their own trucks on regular routes and thereby offering better service, and as good or better prices. We never put a serious effort into servicing these accounts and were not that big of a factor in the service center business in Omaha proper, either. So, to be fair to Howard, he actually did okay with the limited support he was given.

Seemingly, Howard never seemed to tire of driving those Nebraska highways, his "north" run and his "south" run. We knew that he passed by a lot of substantial accounts, and Mike's inside joke was, "He drove straight out, tagged-up in North Platte, and headed for home!"

Howard smoked big, smelly cigars and said he would take an order for "a battleship" if someone would give him one. He didn't get one. He also liked to tell the story of the young Ed Owen, Fred's son, who was working for PVS, and even though very young at that time, let it be known in no uncertain terms that his ambition was to eventually become king of AkSarBen! Howard thought this was really bizarre, but as it turned out many years later, sure enough, Ed realized his dream, and was king.

It was pretty obvious that the Owenses had just used Howard to make that first contact, and then later managed to get him to resign from his position in sales.

Mike and Shorty manned the phones as inside salesmen. I filled in here, too, but I tried to get out and make calls as much as possible. We were learning

house of STEEL

STEEL-INGS

FROM DRAKE-WILLIAMS

REMEMBER WHEN THE ONLY RED MENACE WAS SUNBURN

And have you seen the new Junior Atomic sets? They're for kids who are determined to go places.

* * *

Many an ocean traveler who starts out first class winds up sickened.

* * *

THE NEW BATHING SUITS DON'T KEEP YOU GUESSIN' THEY SHOW A LOT OF SELF EXPRESSION.

* * *

Here's a problem with which we're tormented: What did moths eat before clothes were invented?

* * *

Cards: Where a good deal depends on a good deal.

* * *

Latest Presley hit: "Love me, legal tender!"

* * *

The only full-blooded Americans are mosquitoes.

* * *

MOST LOVE TRIANGLES ARE WRECKTANGLES

* * *

Remember when only Indians ate out of doors?

Things are looking up! As one camel rider said to another camel rider: "It looks like we're over the hump!"

* * *

Regardless of what people will say, a little flattery goes a long, long way.

* * *

It's always fair weather when good fellows get together—but the storm breaks through when they get home.

* * *

REMARKED THE DENTAL PATIENT, WITH GRAVITY: "KEEP YOUR BILL AS SMALL AS THE CAVITY!"

* * *

Young lovers don't know enough to come in out of the moonlight.

* * *

Travel folder: Trip tease.

* * *

Singing in the shower: The bathroom scale.

* * *

A HOT DOG FEEDS THE HAND THAT BITES IT.

* * *

People go on picnics to get tired in a new spot.

AUGUST 1958

"Steelings," the direct mail novelty piece we sent to customers across the state of Nebraka in the 1950s.

pretty fast, and I'm sure we thought we were becoming quite a factor on the local steel scene.

We contracted for a syndicated mailing piece, which we mailed all over the state and to Omaha potential customers, and called it "Steelings from Drake-Williams." Acme Mail Advertising Co. (Warren Youngstrom) handled the mailing, and we used it for quite a number of years. We felt that since Howard, as one man, was spread pretty thin, this gave us a little more exposure and a little help to Howard's efforts. The House of Steel appearing on the piece was our old logo.

Meanwhile, it was 1959, and a young upstart revolutionary in Cuba came from the provinces and overthrew the dictator Batista. He received help from some in this country, plus moral support from Harrison Salisbury of *The New York Times*. Then, Fidel Castro casually announced to the world that his was a communist revolution, he was a communist, and he thumbed his nose at the U.S.

We were entering upon a decade of revolution all over the world. We were paranoid about the Russians, and we were engaged in a space race, as well as an arms race with them. Nuclear war was a very real fear, and we all knew that the Russians also had the power to deliver a bomb virtually anywhere in the country. Children in school had air raid drills, and bomb shelters were sought after, and arguments raged endlessly about what kind of a shelter would afford real safety in case of attack.

We became involved in an oblique way. Other steel companies, including Gate City and Eaton Metal Products (now long defunct), had actual designs for shelters and had taken some orders for home shelters. It was at this time that we were solicited by a fast-talking "salesman type" who wanted prices on a shelter on which he had worked out a rough design. He was intending to sell these practically on a door-to-door basis to the general public. I was reluctant to work with him, doubting his solvency, and not really believing that we needed to build shelters, but I did give him some rough numbers.

This was all I heard from the guy for about a month, and then one day, it happened. One afternoon, an elderly couple walked through our front door and wanted to know if they could pick up their shelter. It seems he had sold them (and collected in full) for his shelter and then told them to stop down at our shop to pick up the beast. They were obviously nice, honest farmers, totally deceived, and had spent the previous day removing all the jars and preserves from their storm cellar at their farm, to make ready for the bomb shelter. They were embarrassed to be the objects of such a scam, and the guy had obviously scared them to death in order to bilk them out of their money.

It was a crazy time. What good would it do to survive in a shelter, if you were to emerge later to a world of total radioactivity? About this time, we had somewhat of a reorganization. We were beginning to make a little money, and we did some house cleaning things within our corporation.

I was elected to be secretary of the corporation, and a resolution, which called for the payment of bonuses of a specified percentage amount and to be determined by shareholder/managers, was adopted. At my urging, the name of the corporation was changed from Drake-Williams-Mount Co. to Drake-Williams Steel, Inc., and this was officially done. The name was registered with the Secretary of State in Lincoln. I argued that the new name would be so much more descriptive of what we were really about, and less confusing to the trade. We decided (rightly or wrongly) to keep the "Drake" name, not wanting to lose the name recognition.

Lest We Forget the Names

Along the way, we were fortunate to be able to hire some really good people for the shop. One of these was Bill Fuhrman. I happened to be

Spiral staircase fabricated for the Omaha Building.

out in the East shop when this old guy came down, asking to see me. We had an order in at the State unemployment office for workers, and Bill had applied at the office. Shorty told him he was too old, but if he wanted to, he could come outside and look me up, but chances were it wouldn't do much good, or some words to that effect.

Bill found me, and so I talked to him, and my first thought was that he was pretty old, also. I think he was about sixty-two, but he told me he was certainly not afraid of work and that he had been a bootmaker for years and worked on a piece-work basis. He was born and raised in the hill country of Missouri and had worked hard all of his life. Something about him told me that this was a good risk for us. His story had the ring of honesty to it, so I told him to go on back to the office and fill out the necessary employment forms.

Thus began the second working career for old Bill Fuhrman. He worked with the yard crew, putting up warehouse orders, and he worked on the rebar crew, cutting thousands of bars, bailing, and shipping tons of rebar day in and day out. We never had a single complaint from the man, and he out-worked every young man that we put with him. He worked on and on, and we could not believe how the man never changed, and never ever showed his age! He finally retired when his superior physical strength finally began to wane. I believe he was around eighty when he finally hung it up!

Another unforgettable in our shop history was Don McAndrews. This little Irishman was a leader among the men and a real mechanic who could build most anything. He had a long and very productive career with this company and was the guy we would rely on to handle the really tough fabrication jobs.

His most memorable contribution among so many would probably be a complicated four- or five-story spiral staircase that he built for the Omaha Building at 16th and Farnam, for the law firm of Kutak Rock. There were other very involved fabrication jobs that Don head-manned, but probably none any more difficult.

At one time, we were moved to establish an annual award for most improved or advanced shop employee, the award being in his name. Don was from a great Omaha family, and his sister was married to Don Leahy, the very capable and well-respected coach and athletic director at both Creighton University and University of Nebraska–Omaha. Don McAndrews was a settling and calming influence on his fellow workers, when at times, during labor negotiations and meetings, there would be a few tempers flaring on the part of the workforce. He was fair-minded and used a great sense of humor to good effect.

Of course, there are many, many others—too many to cite by name. Lots of former employees, as well as current ones, by virtue of being so reliable and steady in their attendance and efforts. They are the heroes of this story. It would be impossible to name them all, but most have accepted total responsibility for their parts in the team effort, which makes the company so successful. And they have learned that cooperation makes the company stronger and able to keep on growing.

It was not always so. Labor negotiations were a yearly fact of life, not looked forward to by us, as management, and probably not the favorite thing of the "committee" either. We always looked on it as unwelcome interference by outsiders (the United Steelworkers of America reps), the first of whom in my experience was a short, fat guy named Floyd Piper. He was argumentative and surly. It took me a while to see through their strategy, which was to try to portray us as anti-labor and unfeeling toward our own people. They were right to the extent that we failed to keep the communication lines open enough.

Art Williams and Larry McMahon sat in also on these meetings. Larry was pretty shrewd about keeping his "cool" during the meetings, but Art was not too insightful. I remember one year, he showed up at the first meeting with a little "sign" which he had clipped from the cover of some magazine or other. It said simply, "We don't want any cheese, we just want out of the trap!" The union was not too amused, but Art certainly was.

In those days, there was a service group in town called Midwest Employers Council, whose mission was to help companies (especially small ones) with employee relations and concerns of all kinds. They were not primarily concerned with union/company negotiations, but they had a manager, John Tate, who was a graduate attorney, and who had had lots of experience with unions and employee/employer negotiations.

When the union first found out that he would be involved, it created quite a stir and plenty of objections. Talks immediately took on harder "edges," and the new man that had replaced Piper, Earl Graham, maintained that Tate was a professional "union buster." Negotiations grew more protracted, and I'm not sure how much good that the MEC really did for us. Once or twice, it got our meetings moved to the offices of the Federal Mediation and Conciliation Services. But we avoided any strikes and did keep the labor contracts in reasonable bounds.

In later years, we initiated multi-year contracts, which was an improvement. We turned to a local labor attorney, Sam Jensen, to help us in our negotiations. Sam was very effective in terms of knowing the fine points of the law, but he was abrasive and almost contemptuous of Earl. At times he seemed to be more interested in all of the legalities, and much less interested in settling with the other side. He took pleasure in irritating their committee. He did succeed in showing the union that their representative was really not able to stand up against our team. I wished that we could spend more time and energy building bridges and solving common problems within our company, and less on finding all the things that tended to divide us.

The story of Frank Britain is an example of solving one of those problems. Frank had been employed by us for several years as a warehouseman and was doing satisfactory work. Trouble began for Frank when he became involved in an accident, and he was charged with driving while intoxicated. As I recall, he got off with a fine and probation from a jail sentence. Unfortunately, this was not the end. Around Christmastime one year, Frank was involved in another accident, and under the same circumstances. There was a fatality in the car that Frank hit, and this time, a prison term was not going to be avoided.

I wrote Frank a message in Lincoln and told him that I was willing to help him under certain circumstances. These were that he would enter into a formal program for alcoholism treatment and that he would agree to continue in this program as a permanent condition of his reemployment.

I wrote letters to the Department in Lincoln and assured them that if he was granted probation of any kind, that he would have employment under these conditions. I visited him in Lincoln, and he assured me that he would straighten up and follow the rules that I had set down for him. Finally, he did obtain his release, some months earlier than his sentence had been set. Frank came back, followed the program, and stayed sober without a single slip.

Frank worked for many years after this and turned into a very valuable and loyal employee. I believe that he never missed a day of work over a span of many years. He proved that rehabilitation was possible, with the right attitude and lots of hard work. I felt especially good because he had truly kept his word to me.

Then there were other guys we remember for different reasons. One was old Tony McAndrews, Don's cousin. Tony, like quite a few of our men, would go up to the corner saloon, the Hickory Inn, for "lunch." One day, he was returning down the hill, and Art happened to be out by the front gate.

"What have you got in that paper sack, Tony?"

"Beer, you damn fool," was Tony's smart-assed answer. And that's exactly what it was!

Good employees were hard to find. We went through dozens trying to get reliable people. If we had an application from someone from out in the country, we hired them on the spot. Always, the ones off the farm or from the small towns were guys that would give you the old hour's work for an hour's pay.

We had a truck driver who injured himself when he backed off the bed of the truck into thin air and hurt an ankle or foot. He was drawing disability until we found him dancing (he loved to dance with the ladies) in the evenings at the Hickory Inn.

We had Christmas parties, after which we needed to bail out one or two later in the evening, when they ran afoul of the law. We learned soon enough that it was not a good idea to serve too much free beer and liquor at those functions.

Replacing Hoody

Business was pretty good, and Drake-Williams Steel was making some headway. George Hoody had retired in 1958, and Art felt that we needed to find a really good man for a replacement. We honored George for his forty years of service with the obligatory gold watch, plus a commemorative banquet at Cantoni's Grill (in the beautiful Marine Room). The room was large enough to hold most of our employees.

George was a good and loyal associate and deserved more than we were able to provide, since we had no pension or retirement plan at the time. I was well aware of that fact, and it was a factor in my later thinking about deferred plans and retirement, and the void that we had in our general employment program. I knew we had a huge weakness and had nothing to offer as inducements to new people.

Additionally, we needed our people to know that we cared about their work and welfare. After we had this party honoring Hoody, Arnold Grabbert came up to me, and he said that he, too, had about forty years with the company. I had no answer, as we did not throw a party for him for his service, and in fact had little to give him in appreciation for these years.

I put this down in my consciousness for the future, and I knew we needed to pay a lot more attention to human relations, as it is termed these days. The profit-sharing plan that we put in later on was too late for Arnold, and quite a few others, unfortunately.

But we needed a replacement for George, so we ran ads, including out-of-town newspapers, and a national publication or two. We found no one. It was somewhat humorous, since we could learn only so much by reading the responses that we received, and we were not going to send anyone on the road to interview, nor pay the fare for any of these respondents to visit us in Omaha. And we were making the basic mistake of trying to find another George Hoody. In other words, a foreman and an experienced layout man. George had been our layout man for plate and sheet metal for all those years, and no one had been trained in his stead.

Finally, the problem was "solved" by Larry. He had located Leroy Shoell, a big, rough character who had worked as a straw boss in the field for a frameless metal building dealer/erector and who knew nothing about layout, nor did he have a bit of experience in shop practice or fabrication. He was energetic, and brash, not conscientious about the details of much of anything, but he got us "by" for a while. He was probably further from being another Hoody than anyone you could possibly find.

We did not find a replacement for Hoody at all, because there just weren't people around with the same combination of skills. Usually, a good layout man was not a good superintendent or foreman. Different skills, technical and communicative. I hired a guy I knew from Trailmobile, and who had run their parts business there, Al Riskowski. Al tried, but he just did not have the background and experience needed, and so we let him go. In fact, we had no one until we gave the job to Duane Shupe. He didn't have all the experience in the world, but he soon learned the necessary skills, including laying out sheet and plate work. He became everything we needed, and more.

Bill King and a Vision of Structurals

Meanwhile, I was continuing to work with Wilson Concrete Co. and knew a lot of their people. Their operations were run by

Don Cerven and, under him, Ron Jackson. These guys were under tremendous pressure at all times from higher ups, including Charles Wilson himself, and the pressure came right on down to all of us. The pressure was so intense, it was not a surprise when both Cerven and Jackson were totally burned out and quit after a few short years.

It was at this time that I met and became acquainted with William F. (Bill) King, who was an engineer on their staff. We had decided that we needed to make a move into the structural fabrication business, and we would certainly require technical help. I decided that Bill was our man and began talking with him about coming on. Bill had had earlier experience with a fabricator in Wichita, and I don't know if he was comfortable coming back to the steel business at that time. I convinced him to jump, and our first problem was where to put him.

We had zero office space, so finally, we decided to clean out the "cellar" under the office building and to finish it off into an office. So there was Bill—down in the "coal hole." We put some kind of a two-way speaker phone down there, so we could "holler" at him whenever we needed to ask him a question, which was often. His office was so small, he had trouble rolling and unrolling full-size blueprints in the space. But Bill did just that, and he did tons and tons of detailing and drafting for us, and was certainly a key player in our growing and learning the structural ropes.

In addition to that, Bill did a lot of basic engineering for our own plant needs and designed our bridge craneways and the buildings that surround them. Bill was under constant pressure from all sides in his job—with deadlines and urgent, rush projects, and demands coming from the shop below and from management above. He was a valuable addition to our firm, and his help was absolutely essential during these years.

In later years, the stress and strain of his job began to tell on Bill, and he came under a lot of criticism from fellow employees at times.

But, all in all, he did a hell of a job for our company at a very critical juncture in our history.

I referred to the fact that we had no one to do layout work, since Hoody was gone, and Leroy Shoell was not capable. So Bill was called upon to solve the problem, and he would make a cardboard or paper pattern, instead of laying out the piece on the steel itself. This was not a great solution, since Bill did not always understand "take-up" and "stretch," which occurred when bending took place. But somehow we got by, and as we began to do more structural work, we were doing less and less plate work anyway.

Once again, we were growing, but not without the growth pains. Finished product was going out the door, and mistakes were coming back almost as fast. I think the great amount of growth in the local economy was saving us to some extent. We needed work space, and we needed tools. We had so little, and so we planned for our first expansion and upgrading, almost the first ever.

Expansion and Modernization

About this time, Cushman Motor Works in Lincoln was going out of the casting business and had been sold to Outboard Marine Corporation. Larry had got wind of this and attended a sale that they had of excess equipment, not needed since they were closing down their foundry operation.

We were planning to build an extension to our West shop, the old brick shop building. The addition would be along the east side and extend from north to south. Larry bid on two cranes that were part of the sale: one a Shaw-Box 5-ton, fairly modern, and the other

1959 erection of steel for addition to the West "brick" shop—the first new construction at the plant in over fifty years.

an ancient and decrepit Whiting—a DC-operated monster. We determined to use the Shaw-Box, and we incorporated it into the new building. The crane had a span of 39 feet and some inches, and so, incredibly, we made the building to these exact dimensions. We apparently did not know or we did not consider lengthening the bridge and remodeling it so that we could squeeze a little more width (and serviceability) into the new structure.

The other crane was put into storage, for possible use at a later date. This we did, much later, and this crane was lengthened, and a lot of money was foolishly expended to remodel and update this old antique. This was a huge mistake on my part, but that is another story, for a later time.

Bill King put together the design for the new shop, designated as "D-59" by Art. He designed the A-Frames and the plate girders that would span to the old building. The girders were made from a bunch of old UM plates that had come from Larry's Nu-Steel inventory, part of the steel inventory that he used to buy into the company in 1955. We went to great lengths to save a few bucks and possibly spent the savings of material on extra labor, but we knew we had to scrimp to get as much mileage as we could out of every dollar.

We determined that we were going to need around $165,000 for this expansion project, and this included two major pieces of equipment: a press brake and a plate shear. This was a colossal sum of money for us, and for Art, it was the whole world. Growing up and knowing what it was like to live through the Great Depression, his generation was geared to "pay as you go" and was just plain scared of debt.

The money was obtained on a term loan from our regular banking source, the First National. A note was drawn up, and conditions were outlined, such as a prohibition against the payment of any dividends without the prior consent of the bank. Restrictions on salaries and other measures were included. There was no negotiation over this instrument, and no other source was sought for a competitive offer of funds. We were naïve about such things and instead felt lucky to be able to obtain any sort of loan, which we considered a "jumbo."

Art was in charge of the specifying and purchasing of the major pieces of equipment, and he looked at offers from Cincinnati Tool and Pacific Machine & Tool. Hydraulic shears and brakes were brand new, and he was intrigued by the idea of "air" bending. The applications that he had in mind were for bending the plate lips for our line of dragline buckets ("Omaha"), which we had been having to do using heating torches. Also, he wanted the brake for certain heavy plate work that we did for American Smelting & Refining.

Air bending was the term for accomplishing bending without bottoming the knife die into the bottom die, and which could not be done with a mechanical brake. The hydraulic brake had certain

Roofing the 1959 West shop addition. The new 12-foot-wide Pacific plate shear can be seen at the far left, waiting to be installed in the new shop.

advantages—for example, it was slower than the mechanicals—but no one gave him any opposition to this idea. We were impressed by the notion that hydraulics could not be "overloaded" and damaged like mechanicals, by trying to bend or shear something overly thick or beyond the machine's capacity. The gearing of a mechanical could be damaged; whereas, the hydraulic machines simply stop, then recycle, without any harm to the equipment itself. The Pacific brake may be seen under the tarp on the extreme left in the picture, waiting to be put in service.

We liked Cincinnati's equipment best, but Pacific represented quite a price advantage. Art finally chose the Cincinnati brake paired with the Pacific shear. As I recall, the brake was close to $40,000 and the shear about $33,000, and these were immense sums to us. We were eager for delivery once the orders were placed.

Harris & Meyer crane setting the 5-ton bridge crane for the new shop addition in 1959. This was the first and only overhead traveling crane D-W-S had at that time.

We took bids on the new building, and Ernie Vrana (of Chas. Vrana & Son Construction Co.) was the low bidder. These were heady days when construction started—as the first dirt was turned and the old lean-to wooden structure was torn down. This old structure was used to house our boiler and condenser tubing, and we had failed to move this extensive inventory out of the way for Vrana to tear out the old structure. So Ernie tied into the job of moving all the materials, which was something he had not counted on in his bid. Extra charges right away, and expenses we had not counted on.

Meanwhile, the cranes arrived on gondola cars from Lincoln, and we hired Jerry Smith (of Valley Steel Erection) to pick them out of the cars and offload them in our yard. Jerry was a one-man operator who handled the crane himself. He had his small yard out on 60th and Valley—a site that was bought out from under him for the construction

Interior of the new shop in 1959. New plate shear (left) and new press brake can be seen.

of Interstate 80, which was being laid out and built at that time. He was a tough little guy and fearless at the controls of his small rubber-tired crane. Larry insisted on calling him "Valley" Smith instead of Jerry. Of course, Larry had his own lingo and pet names for lots of stuff. He might just call you Robespierre or Sylvester, or some other silly name, any given morning.

The bridge cranes turned out to be almost beyond the capacity of his little crane, and watching him snatching them out of the cars was a genuine thrill. He came within inches of flipping his crane about a dozen times, but finally got the job done. Now we had the cranes, and the next thing to be done was to figure out how to set them in the new building. We had never done this, and Ernie had never done it either, so we hired Harris & Meyer, the steel erectors, to set the crane in place.

Practically everything in the way of tasks that Vrana took on was something that was not shown on our building "plans," such as the massive foundations for the shear and brake, and the dismantling of the old locker room and bathrooms, and filling in of the hole that resulted under the floor of the west building. It was really a good thing for our "cause" to have a guy like Ernie doing the job, as he had a personal interest in our firm and in us as individuals. Ernie not only did all these "extra" tasks, but did them in a hell of a hurry and gave us the finished structure in record time.

Later, when we were struggling to get a toehold in the structural business, Ernie gave us a couple of our first jobs. He was a hard-driving man, a "workaholic" but a very good friend to Drake-Williams Steel, as well as to Mike and me.

We made a lot of mistakes in the new facilities, but because we made some moves and improvements and got some crane capacity, plus nice new tools, the net result was good. I wanted the new building to be back along the south boundary of our property, running east and west, because I felt that we would be too restricted at the north end, running into Hickory Street. But just the fact that we were on the move, and progressing, was enough for me. I was anxious to go out and get customers and to fill the new facility with business.

But despite the new building and tools, we still had virtually no structural fabricating equipment. We did buy a Do-All band saw with a capacity of 18 inches, so we didn't have to flame cut all of our sections. We had some single-end punches, and a Pels ironworker. We acquired some magnetic base portable drills, which were a big help. The only beam punch on the market was a Beatty, and it was pretty slick, but not good for ones and twos, only lots of duplicate pieces. But a Beatty was not in our price range. The Whitney and Peddinghouse beamlines had not yet come on the scene.

Whitney did have a line of portable hydraulic punches in "C" frames that they were hawking, and I had the idea of setting some of these up along a conveyor, which was exactly the idea that Whitney finally brought out in their first beamlines. In fact, later on, I gave Hoody some rough sketches and catalog information from Whitney and asked him to design a "line." He came back to me about three weeks or so later and said that it was just too much for him to figure out. George was up there in years and probably not able to concentrate really well.

I didn't take it to anyone else at that time, but it wasn't too long before Whitney brought out their beamline. I was amazed and really excited when I saw it the first time and was determined to try to get one at some point. But this was not to happen for quite some time.

Competition Moves In

Omaha must have looked good to outsiders along about the early 1960s in terms of the amount of business activity and growth. One day, a guy by the name of Al Ray stopped by to make a courtesy call. He had been a friend of Larry's and of some of the Gate City people out in their Boise, Idaho, plant. He announced that his company, Midway Steel, would open a facility over in Council Bluffs. They would be strictly in the "warehouse" business (which was just becoming known as "steel service center" business) and would not be fabricators. It was Al's idea that we could cooperate, and they could send us fabrication, and we could give them some of our warehouse purchases.

This did work to our advantage for several years, and we gave them a fair amount of our business. Mike kept them at arm's length and did a great job of making sure we did not come out on the short end of any "deals" that we might make with them. Then one day, we discovered one of their sources of supply was Robinson Iron and Metal of Des Moines, so Mike began to use them in lieu of Midway, for certain items. Finally, Midway sort of petered out, and they quietly closed their doors and faded away.

Mike, by this time, had become a very important reason for our success. He was a really good buyer, and without the supply, there is nothing to sell. We worked from hand to mouth, and with a limited availability of cash. Mike remembers one day when Duane Shupe, our burner, was working in the rear of our property, way away from the office. Later, some wood cribbing had caught fire and was getting out of hand. Mike called the fire department and asked them to send just a small unit down with a couple of fire extinguishers to put out the fire. Minutes later, all hell broke loose as several large fire trucks screamed into the yard and began stretching hoses across the property. Mike was a little embarrassed over this one.

I don't know exactly how much we have entered into the equation, but I wonder if it is any accident that, over the years, we have seen quite a list of substantial outfits close their doors and quit the scene. These would include the aforesaid Midway, plus Omaha Steel; Gate City Steel; Northwestern Steel and Supply, these in our immediate vicinity, and also Lincoln Steel; St. Joseph Structural Steel; Des Moines Steel; and Christopher Steel, of Wichita. I know that we gave them all the competition that they could handle (and maybe just a little too much).

Demise of an Old Name in Dragline Buckets

All this time, we continued to limp along with the bucket business—the Omaha brand. Our biggest competitor was Hendrix, from Mansfield, Louisiana. Art told us that they had a big advantage in that they had their own foundry and could cast their own bucket lips out of manganese steel—very strong and very wear resistant. The other advantage was their location in the South, near the areas of the greatest employment of this kind of equipment.

It was apparent to me that they had a big lead on us, even though we were out in the market long before they came along. The first in the industry was Page, out of Chicago, and then we came along, possibly in the 1920s. I don't know who actually designed our line of buckets. It was probably sort of a joint effort between ourselves and some of the river and bridge contractors at the time.

The contractors were naturally interested in capacity, which made for efficiency in their operations. One of these contractors at the time was named Peterson, and he had the nickname of "Big Wagon," thus "Big Wagon Peterson." He apparently helped us in the initial designs and encouraged us to build bigger and bigger buckets.

We had been making our product for several years, when suddenly we were sued by the Page people in Chicago, claiming patent infringement. We began paying them a small royalty on each bucket, as a settlement, since the suit never went to court. But it was never clear to me how this came about, since it was claimed that there never actually were any design patents on a dragline bucket.

The bucket business was Art's favorite, and he dreamed of overtaking Hendrix in the field and really recovering that part of our business. In fact, this was the reason that Art's friends nicknamed him "Buckets."

I made several trips to the annual convention of the A.E.D. (Associated Equipment Dealers) convention in Chicago. It was always held at the old Conrad Hilton Hotel in the Loop, which I believe was the largest hotel in the world at the time. These trips were not too productive, although we did make a few "contacts" that later produced a few sales. It was on one of these trips, as we were just getting ready to start the train trip home, that we heard about the Starkweather murder rampage in progress. Scary stuff, and I was particularly concerned about my wife Mary Jean's parents, who lived right there in Lincoln. Starkweather held the state and the whole region in terror for about a week before they captured him and his girlfriend, Carol Fugate, in Wyoming.

Calling on the trade, it was apparent to me right from the start that our product was not much of an item for the dealer. He was interested in the crane sale, which, of course, was the big ticket item. The tail didn't wag the dog, and the dealer wasn't going to spend a lot of time on our stuff, because the money was simply not there. He wasn't interested in trying to sway a customer toward the Omaha, if there was any chance of losing the big sale of the crane in the bargain. By the same token, the user didn't really have a big preference in the matter, either. His attention was on the crane also and how it performed on the job. So, I saw it as pretty frustrating, and a hard sell.

Further complicating and frustrating the picture was the fact that we were not really making any margin on the bucket orders that we had. We knew that we needed to make them in some kind of real quantity to realize efficiencies, but we just did not have the volume. The lone bright spot in an otherwise gloomy financial picture was the parts business. We did have quite a few buckets out in the field and, at first, got quite a few parts orders.

This process is laughable by today's standards of overnight shipping and Internet communications. We received our orders by (snail) mail, from a dealer, so no doubt several days had already elapsed from the time the order was placed by the contractor. We would process the order probably the same day and ship the item a day or two later, assuming everything was in stock. The order would be picked up by a local trucker and would go to his Omaha terminal, sit on his dock, and maybe go forward in another day or two, so probably the contractor would get his parts maybe two to three weeks from the date of order.

But the best parts business was the government (Army or Navy) business. During World War II and to a lesser extent the Korean War, we sold quite a few buckets to the services. Somehow, out of the myriad offices that constituted the Defense Construction Supply Agency would come an inquiry for parts. Not for ones and twos, but for many, maybe hundreds, of things like sockets, clevises, cable wedges, and trunnions, and we knew, of course, that we were the only source.

Naturally, we would quote full list price and back would come the orders, like clockwork. We knew also that these parts were not being put to any immediate use, and probably would never be utilized. Some agency, somewhere, just needed to use up their fiscal year budget, in order that they not lose the funds in the next fiscal cycle.

This agency was in Columbus, Ohio. Art was there once and told me that the warehouses there were enormous and absolutely brimming with every kind of part and tool and piece of equipment that you could ever imagine. On a few occasions, we would also get lists of surplus parts that the government was offering for sale. Also, more than once, we would be buying some of our parts at junk prices, on the one hand, and selling them right back to the services, with the other hand. Peanuts in dollar amounts, but making you realize what huge duplication and waste was going on with other transactions, dealing in much bigger and more numerous parts.

The bucket business continued to dwindle, and after 1962 or 1963, we did nothing to promote it. We made so few actual buckets that our shop people were forgetting how to build them. We did make an attempt to sell the business, together with all the patterns and inventory of parts, to Owen Bucket Co. of Cleveland, Ohio. They turned us down, and after that, the Omaha dragline bucket just died a slow, natural death.

Around this time, Art began to spend less and less time at the office and shops. One day, however, after regular working hours, and a few months after the new building and expansion had been finished, Art called Mike and myself into his back office. He kept a little refrigerator for ice back there, and so had himself a little portable bar by his desk. He was having a little scotch and water, and he wanted to show us something. It was a notice for an installment on the note from the First National Bank, and he announced that it was due in a very few days, and we just didn't have the money to pay it.

"Boys" he said, "we're screwed, we are just screwed!"

Mike and I just looked our amazement at each other and tried to keep from laughing. We knew a little better than he did the amount of business we were doing and knew that we were not hung out to dry—not just yet! Mike and I used that over the years, as an "insider" when we wanted to sort of put things in perspective and remind each other of where we had been.

But, in his defense, Art and, for that matter, his eccentric older brother Walt, had lived through the deepest depression that this country had ever weathered, and they had seen this through in their younger days, while watching many, many local and other businesses go down the drain. They had learned to avoid debt at all costs and to owe no man. Do without first, before you borrow from anyone, least of all a bank! They had seen the banks in their repo modes, and they wanted no part of being on the wrong end of the collection process.

After that, he was gone more and more. If he did come down, he left right after lunch and didn't return. It seems he was practicing for a retirement. When he was gone, it seemed Larry would be more in evidence. Larry became more friendly to Mike and myself, and we joked a lot and got along really well. Larry was a genuine character, and we played little games, like trying to find him in a mistake. He liked to do math in his head and could add a column of figures like crazy. We would bang away on our calculators, and Larry would just sit and figure, his only aid being a common wooden pencil. Then he would dare us to find a mistake on his pages.

Fun and Games

One day Mike asked him a question about his estimating method. "Larry, how come you charge 50 cents each for bolts?"

"Well," he said, "got to. King leaves out about half of them, so we have to make it up somehow!"

Larry had a birthday sometime during the summer months. Mike and I came up with a little plan. I wrote five short poems, all about Larry and his foibles. One was kidding him about "little Audra," another about falling in the river, which he did once when he was trying to retrieve his duck decoys, and so forth. We read them all to Western Union and instructed them to call each of them in, strictly to Larry, no other, at each hour and on the hour, starting about ten o'clock.

The first couple were pretty baffling to Larry, but then he caught on. He tried to act unimpressed and not at all amused at first. I guess he thought we were making a little too much fun at his expense. Later, Audra told us how much he laughed about it, and how much he actually appreciated the whole caper. I found out later that he told Audra that he actually felt as if we were the sons he never had.

Larry would be the most surprised and delighted guy in the whole world if he were still alive and could see what we have done with his old Gate City property and the changes and vast improvements that have been made. He simply would not believe it.

Still operating out of our tiny office, we knew we were not going to be able to use a single extra body without some space. We had an old garage building to the west of the office, which was really an extension of the West brick shop. We added slightly to the north end of it and put in a large plateglass window and glass door and remodeled the place with paneling and carpet, and a heating and a/c unit in the rear, and we had ourselves some space for King and another draftsman to work. It was Rube Goldberg enough, but we sure needed all the room we could find.

Back in those days, Mike and I found a novel way to pass the time in the dreary winter days on Saturday afternoons. We brought a couple of our hunting buddies down with us and did some great pass shooting of the pigeons that flew along the UP tracks. They were attracted by the spilled grain along the line and were always plentiful, and not too high off the ground. We did this quite a few Saturdays, hoping that no cops would come along and see us.

Mike had a friend, Hughie Harger, who owned Ideal Roofing Co., our neighbor to the east. He was aware of our Saturday afternoon "hunts," so he pulled a little stunt on Mike. He had a friend on the police force call Mike early Monday morning.

"We have a report of illegal discharging of firearms within the city limits at this location on last Saturday. What can you tell us about this?"

Hughie got him good, and Mike fell for the whole thing!

Back in 1960 or 1961, I discovered a gold mine in the form of the Western Electric project. I called on a company called Newberry Electric and was able to pick up some miscellaneous fabrication. Then I found Commercial Contracting, a company that originated in Detroit, and it was like the mother lode. They had the contract for the moving in and installing of all the W.E. equipment in the 30 building, which was the first and the westernmost of the two main structures. Commercial Contracting dated from the years just after World War II, and they became specialists in moving and mothballing equipment of all sorts, equipment that had been used in large defense plants around the country.

They needed miles of pipe guards, special angle table legs, and all sorts of miscellaneous steel. Time was always of the essence, and service was the key thing. If I wasn't out there every day, I was on the phone with their office constantly. For the small company that we were, this was the big time. There were even some structural requirements from time to time. We feasted. We ran Gate City out because we were giving them service and attention, and price was usually not a problem.

My buddy was a guy named Rudy Pitvorec, and I learned a lot from him, mainly about the workings of large corporations such as Western. The amount of waste and the extravagance was mind boggling to me. Used to squeezing all the pennies and dimes, it was almost more than I could comprehend. About this time, Rudy told me about his boss, from Detroit. He asked him if Western was bucking back any of his price proposals, and that if they weren't, he obviously was not charging them enough. They were on the site and were the only real bidders to do the various parts of the work, and the "boss" wanted to maximize the revenue, and the profit as well.

We had one old International flat bed truck to our name, and maybe a pickup truck, and that was it! I designed a logo for us (the House of Steel) and came down to the plant on a Saturday and took black paint and lettered it on the sides of the truck. I was so anxious for us to be in the warehouse and fabricating business, and I wanted us to look like we meant business and were a forward-looking upbeat outfit. Later, I designed the trademark or logo that we still use today. It has served us well.

We hoped this gold mine would last forever. Naturally, it didn't, but soon it was time to start the 50 building, called the "Cable" building. This time Commercial was not able to keep the contract and were beaten out by Huber, Hunt and Nichols, out of Indianapolis.

Commercial, however, was able to take another Western project, this time in Oklahoma City, where they were building a facility almost as large as the Omaha works.

Larry had met these guys somehow, and we were again on the ground floor, furnishing all kinds of miscellaneous items for this newer building. This time our contact was a guy named Lefty, and he was tough. We did a lot of business, but not without taking a lot of heat with it.

Lefty would call for Larry at our office, and Larry's usual excuse? "Paint's not dry, Lefty." This did not hold up very long, and soon Lefty would be down at our office and in the shop, demanding to see the item in question. Things got very interesting at times, but we somehow managed to satisfy HH&N, and I think we still got the lion's share of the business.

Later still, Huber, Hunt and Nichols got a contract also down in Oklahoma City, this time for the general construction of one of the major buildings. Later, we got a job with them, building lightweight trusses for a crane system. Lefty and the boys, remembering the great steakhouses and Italian bread here, had us throwing the loaves of Rotella's bread that they loved right on the trucks with the trusses. Crazy guys and crazy days!

All this business was good, but we still were crawling along, not able to be any kind of factor in the structural steel business. Mike and I wanted to grow and to expand, and we were casting envious glances to our west, where the property, which was formerly owned by Mobil Oil, was now owned by Watson Bros. Transportation. We tried to talk to Larry about buying this, but he just stalled around and evaded any direct answer.

We tried to be a comprehensive warehouse (or steel service center in the new vernacular) but we lacked sheets and galvanized, and although we traveled Howard Coe out in the state, we really did not have a "full" wagon. We had a salesman, Joe Gurnett, on the streets also, but neither of them got the real backing that they should have had. Joe never got any real sales volume, for whatever reason, and so Larry fired him suddenly one Saturday morning. Larry thought he was lazy, and he was doing him a favor to let him go.

We had some interesting special jobs along the way. Along the Platte were the dehy plants. These were the boys that were involved in dehydrating alfalfa and making it into cattle feed. We made a couple of dryers, and one in particular for Otto Albers in Wisner, Nebraska, was a real project. It was 12 feet in diameter, but was closer to 13 with the tracks on each end. It was probably 24 feet in length.

For a time, we thought we had goofed, like the old story of building the boat in your basement and not being able to get it out. We had to map out a special route to Wisner and provide an escort all the way. Nevertheless, a few wires were downed, and some tree branches got a quick trimming. But the dryer worked well, and the job was a success. I guess we imagined making lots of these, but the business did not really develop into that much.

We seemed always to be taking on one-of-a-kind fabrication jobs, and always with its own particular problems. We tried our hand at making concrete forms for double-tee slab pieces. These had to be extremely straight and accurate. We made them for Wilson Concrete and for Ideal Cement Stone Co. I will give our shop people credit, as they always tried their best to do these difficult jobs and always succeeded in satisfying the customer.

But we were making some headway in the structural business, despite the lack of equipment and facility. Larry had a good relationship with Simon Korshoj of Blair. They called him the "King of Blair," and he was quite a guy. He had a reputation as a general contractor for taking the hard and messy jobs—the ones that no one else wanted. He usually made them work out, and he was a pretty tough dealer.

"Topping out" steel at Borup Activities Building, Dana College in Blair, Nebraska, around Christmas 1965.

The other thing was that Korshoj's man in his Omaha office was Del Chapin, and Del just happened to be Larry's brother-in-law. Consequently, Larry was able to get quite a bit of "feedback" on letting days, and we used that information to very good effect. Simon Korshoj got nearly all the work at Dana College there in Blair, with his local connections and buddies.

And so, we had a large job going in the autumn of 1965 with Simon—the Borup Student Activities building. It had a large domed roof and was a tricky fabrication problem. Sloping roof beams angled in all around the perimeter to a compression ring in the center. Simon told us later that the whole thing went together without them having to use a cutting torch even once. This was a feather in our hat, since we were used to back charges and screw-ups on lots of our jobs, and this one was a "dilly" to make fit! At 160 tons, it was the largest structural job we had attempted at that point in time.

All this time, we were learning. Mike and I were learning how to estimate all kinds of work, including structural. We were working out systems and trying to get away from "pound pricing," which was extremely risky, and not tied in any way to manhours. Little by little, we were working hours and labor into the estimate, separate from material.

A Halloween Horror Story

I am jumping ahead of my story a little bit, to a day in October, the 28th, 1965. I was invited to a dinner given by the Omaha National Bank for promotional purposes and was at the new Kiewit Plaza building. Mike was not with me. Larry had taken the day off to escort some mill people from Kokomo, Indiana, out to a duck blind he shared with Randy Drahota, his friend and dentist.

I don't remember much of anything about the dinner, but I will never forget what happened right afterward, as we were visiting and waiting in the hallway for the elevators. Someone approached and called me to the phone. It was Mary Jean and she told me the bad news. She had a report that Larry had been in a very bad accident out on West Dodge Street and had been instantly killed in the wreck. He was alone in the car, except for some kind of a little animal he had caught that was in a box in the back seat. This may have been the distracting element in the thing, but the fact was, he drove right under the rear end of a cattle truck that was waiting to make a left turn at Skyline Drive. I imagine that the animal was meant to be a joke of some kind to be pulled on Mike and me.

I went home and got Mary Jean, and we drove out to the McMahons on 90th Street. Friends John and Bev Sinovich were there, as well as Sally, the daughter, and a few others, trying to console Audra. Her worst nightmare and biggest fears had been realized, and Larry was gone in a snap of the fingers.

Meanwhile, Art was AWOL. We just continued to issue his weekly checks, and Virginia would bank them for him. If Larry, who felt that they were in effect partners, cared about this, he never spoke about it to Mike or me. Art was just unofficially retired. I'm sure his intentions were to get back in the routine, but he never did. Even after Larry's death in the accident, he was not jarred enough to return. The accident shocked the hell out of everyone, and the next day at work, I was acting president, ready or not!

There were some loose ends as far as the shares in the company were concerned. Many years prior to this, in the late 1940s, when Verne Reisser had left the company, there was some unfinished business. He had left the company, taking the Elegrader parts and equipment with him, and a cash deal had been made, but he had retained twenty-four shares of our capital stock.

From time to time after that, Verne would call or write, usually around the first of the year, and ask some general questions about the health and prospects for the company. I'm sure he thought the company might be going belly up just about any time. On the same day as the funeral, Verne stopped by and again inquired about the business and made it clear that he wanted to sell his shares, and he expected us to buy them. It's the old story of the body not even grown cold, and here come the vultures.

I'm positive that Verne saw the company in the hands of us "kids," and he decided that he had better see what he could salvage out of the old place. The whole idea of him sticking his nose in the door on the very day that we buried Larry really burned me up.

I told Verne that he would just have to be patient, that we didn't have a lot of resources, and we had plenty of problems to face, what with Larry's death in the face of the agreement that obligated us to buy his share of the company. The irony was that Larry was killed on October 28, which happened to be exactly the final business day of our fiscal year.

Larry never lived to learn of it, but the big news for all of us was that we had finally reached a sales goal for which we had been striving forever, it seemed. For the first time ever, we had achieved sales of slightly over $1 million! Larry knew we were having a pretty good year, but we were really sorry that he never got to celebrate that goal with the rest of us.

We were faced with a sizable obligation for purchase of his interests under the buy/sell agreement dating back to 1956. The agreement provided for the total purchase of stock by either party—Art or Larry—at the time. In 1958, I demanded that this be changed so that it only applied to Larry's death, and not in the case of Art's death. I did not want Mike and myself to knock ourselves out building the company, only to have it wind up 100 percent in Larry's hands if something were to happen to Art.

We used "book value" as determined by our annual audit as the basis for valuation of Larry's eighty-two shares of stock. The agreement did not specify any particular method for an evaluation of the shares, and we had used book value as a basis on several occasions in the past, so we did not contemplate any other method of evaluation. We gave Audra McMahon a cashier's check for $134,374, plus another check for his share of the bonus for the 1965 year of $24,421. We thought this was a great pile of money at the time. For us, it was big money. Most of that was borrowed, and we had to work it out and pay it back. We caused the company to retire Larry's stock as treasury stock.

Sally McMahon, Larry's daughter, had asked Simon Korshoj to counsel her and her mother, and Simon tried to argue that the stock should be worth more money, and for a while gave us a little flak, but eventually, everyone seemed satisfied that the settlement was fair and equitable. Simon had very little to get his teeth into, because the numbers were right, and the agreement did not really leave much of anything to chance.

But Simon would call me on the phone and want to talk by the hour or so. He was always asking me to come up and look at some of his prize junk steel—beams and miscellaneous material that he had salvaged from some county bridge that he had replaced.

He was a genuine character, and he ran his business like he really was the "King of Blair."

Not long after this, we cleaned up the stock book by buying back the Verne Reisser shares for $1,367 per share, which was less than the McMahon shares, but we had no agreement and felt no real obligation to Reisser. We felt good. Now the slate was clean, and only members of the Williams family had stock.

Mike and I got along really well, were almost always in agreement, and now we could start making plans and see what we could do "on our own." Only one thing, though. Art still owned his shares, and we needed some way to set a value on his holdings so that we were not working against our own interests, and building value for an owner who was in a retired mode, and not contributing anything to operations.

Reorganization and a New Thrust to the Future

A new plan was worked out for an exchange of Art's shares of common stock in the company for preferred shares. This pegged the value at the current value and provided for a dividend, which had to be paid before any other distribution could be made, and was cumulative if any dividend was missed in any year or years. Since he was retired and was no longer contributing to the growth of the company, the price of the preferred shares would not appreciate, but remain fixed. That meant that, likewise, it would not decrease in any year or years that suffered losses but would continue to pay the same dividend.

The idea of using preferred stock came from Carl Mammel, my friend and adviser, and was a nearly perfect solution for our situation. It worked very well for our purposes, setting a value that was static, and assigning the risks/rewards to the common holders. It provided current income for Art, which served his needs well.

I had long wanted to utilize a profit-sharing plan, and Carl was helpful here also. He had a number of prototype plans, and we tailored one for our salaried employees and put it into place in October 1966. Of all the things I did as president of D-W-S, this is probably the one that made me the most proud. It made a difference, and while not a panacea for all of our problems at that time, it served to give our people a sense of ownership and pride, and I think they began to realize that each person could make a difference, and that they could have a share in the future success of the company.

It was a sort of "growing up" process for the company. We tried to call attention to it and to emphasize its value whenever we could. But it really only showed results once a year, and then it was an account on a piece of paper that promised dollars off at some distant, future time. Our people appreciated it, but it remained for later, when another incentive plan was put in place, to really capture the enthusiasm of the group.

Here is the formal announcement letter that was circulated to all employees on December 20, 1966:

Announcing

DRAKE-WILLIAMS STEEL, INC.,
PROFIT SHARING PLAN

To All Salaried Employees:

The company is pleased to announce the adoption of a Profit Sharing Plan for the benefit of its employees. You will be eligible for the plan if you were a full time salaried employee on October 31, 1966.

The plan is primarily designed to provide financial security for you and your family at retirement. Benefits will be payable, however, in the event of death prior to retirement, or in the event of total and permanent disability.

The plan will be completely financed through contributions made by the company and will depend entirely on profits. Monies will be invested in a Trust Fund for the benefit of the employees.

We emphasize that the plan and its success depend entirely on company profits. It can be a very successful plan and pay handsome rewards to all. It is designed to have advantages for both employee and the company, for example:

CORPORATION -Will enjoy tax advantages.
-Hopes to gain by better motivating each employee.

EMPLOYEE -Gains a share in profits.
-Saves the money, since the payments are held in trust for distribution at retirement, termination, or death.
-Enjoys a tax advantage, since funds paid out at retirement are taxed only as capital gains (Maximum rate 25%) rather than as ordinary income.

Neither the employee nor the company are taxed when contributions are made into the plan.

In the near future, you will receive a copy of the plan and a statement showing the amount that has been allocated to your account.

Hugh W. Williams, President
December 20, 1966

This was a plan that provided deferred dollars, and thus took care of people's future needs and was great motivation for the long term. This first profit-sharing plan was a start, and it told people that their efforts were appreciated, and they weren't just working for a salary alone.

Every chance I got, I tried to preach that we should not think in terms of just "dividing the pie," but we should instead concentrate on baking "a bigger pie." Not a zero sum game, in other words, but a dynamic, growing fund to use fancier terms. Still, too many employees did not look that far ahead, and their current needs and wants were not being met. It was a start and a step toward winning the trust and loyalty of our salaried people.

I wanted to include the shop in our plan, or to install a separate one for the rest of our people. The big hurdle and hang-up was the union. We were not at liberty to just put in the plan, because our contract agreement plainly stated that the union was the sole bargaining agent for these employees, for anything that pertained to wages and hours, and compensation of any kind.

I racked my brain to try to think of a way around this. We were not comfortable to suggest this in the context of any negotiations with the union, which we had periodically, of course. We let the shop people know about the salaried plan and tried to put out the word that we could have a companywide plan, if it were not for the obstacle of the union.

We really got no response to this, but at least it had the benefit of putting the idea into a few heads. We knew that if we installed a profit-sharing plan as part of negotiations, there was no telling where it would lead. We knew that the union would try to take credit for the whole plan if we did this, and that they would try to get a hand in the determination of benefits, and would want to be shown the books and the financials, and have a say in the dollar amounts being deposited in the plan.

We knew that the major benefit for the company was the creation of incentives to spur cooperation and efforts from the shops, and that it could be used to bring these employees closer to the management group. It would work at cross purposes to the wedge that we felt that the union outsiders always tried to drive between us.

It would be about five years before we were able to inaugurate the shop plan. In the meantime, we knew that some of our employees had voiced their desire to be included, and that they felt left out the way things were. So, by the time it did happen, I think there were many who were really ready for the idea. It turned out that our fears about the union trying to co-opt the plan were baseless. They never tried to interfere in the operations of the plan.

The preferred stock was voting, but Art did not somehow understand this. He wanted a say in the company and did not approve of much of the expansion that we were undergoing. He was sure that we were going to break the company, and he told some of our employees this many times.

Gordon Cizek was one, and Gordie tried to be diplomatic and did not want to disagree with him. But Gordon knew that what we were doing was not only right, but probably not really aggressive enough. Art did not say as much to Mike or myself, but he did say to several others that he was extremely proud of how well we were doing with the company. Somehow, even though we really wanted to hear it directly from him, he found it too hard to say the words.

I think anyone who had lived through the Depression years like Art had would be inclined to be overly cautious and pessimistic. Probably if we erred, however, it was on the side of caution.

Only a couple of days before Art died, I was visiting with the folks in the evening, and I told him that we had just purchased part of the Omsteel plant at 48th and Leavenworth. He just looked at me and kind of sighed. I knew that he thought I was quite insane, and that this could be the "final straw," but he did not say much. He just sighed and made no protest at all.

After Larry was killed, and after we were over the initial shock, Mike and I began to feel our oats. We knew that the year 1965 had been a good one, and now we were pretty confident about the future and that we were making pretty good money.

Ron Parks Comes Along

We wanted to add some management and support people. We interviewed Ron Parks and subsequently hired him on December 13, 1965. Ron still has the handwritten "agreement" that we made with him at the time. Of course, this was Ron's idea, him being the organized and focused guy that he has always been. He

Hugh, Arthur, and Mike Williams at site of Topps at the new shopping center on West Dodge Street in 1969.

had come to us from Continental Can Co. and had no formal office or administrative training.

We had him start in the shop, where we would move him around in order to get as good an understanding as possible of operations. The deal was that he would start at $2.30 per hour and would spend from six to twelve weeks in the shop at that rate. Moving to the office, he would be paid $450 per month, and he would get one week's vacation after one year, and two whole weeks after four years.

We didn't know what a "gem" that we had in Ron right away. What we did find out was that he had an insatiable appetite for information and facts. He questioned everything that he was told and asked his questions incessantly. We played games with him after he came into the office. He was learning steel and steel sizes. Mike would call out "2 x 2 x ¼ angle" and he would answer "3.19," which was the weight per foot.

Pretty soon, he had all the strip, flats, and angles memorized, plus all the weights of plate and most of the beams and channels. He soon learned all about our customers and all about purchasing. He and Mike got along famously and had a lot of mutual respect for each other.

This did not justify some of the stunts that Mike pulled on Ron. One day, Mike found some limburger cheese somewhere, and when Ron was out of the office temporarily, he unscrewed the speaker end of his phone and put in a generous amount. That wasn't enough, though, for Mike. We had some steel office chairs with leatherette arms, and Mike took straight pins and, working furtively, managed to embed them in both of the arms of the chair so that they would catch Ron at about the wrists. When they had the desired effect, Mike gave himself away immediately by laughing and crowing across the desk from Ron.

It was obvious that Ron was someone special and that he had a very analytic mind to go along with his curiosity. He became a very good estimator in short order and quickly learned the capabilities of all of our equipment, as well as our people. He did not mind working on Saturday mornings, or coming in early if it was needed. Likewise, he knew our customers and what their needs were. He always took care of business.

Mike made up names for Ron—"Poopsie" and "Chief." Ron railed at Mike about all these "outrages," but he laughed just as hard about them as we did! I knew pretty soon that Ron was going to outgrow the job that we had for him and that he would be moving on soon. He knew that he wanted a business all his own. He was just about ready to run one.

Ron came to me one day with a plan that he had been working on. His idea was to start a galvanizing facility in Omaha. It just so happened that I had done a little bit of research myself on the idea, because we did seem to get a lot of calls for hot dip galvanizing. I had talked to some of the platers around town and had done some rough figuring. I concluded that it would not be a good venture. I told Ron this, and he was satisfied for a while. But it was not long before he came up with still another idea.

Ron Parks Chases His Dream

Next, Ron told me he had an opportunity to buy Millard Sheet Metal of Millard, Nebraska. He had been talking to Elmer Johnson, the owner, and had roughed out a plan. I looked at his plan and quickly told him that I couldn't find any major fault with it, but that naturally, it was not without risk. He was determined, and I could see that, so I told him to go ahead and do it.

We hated to see him go, but it was absolutely bound to happen, sooner rather than later. Of course, it didn't take Ron any time at all to explode that company, pay off the owner, and then rename and relocate the business to 108th Street. Millard Manufacturing has been a huge success, and Ron has become an outstanding entrepreneur and businessman. His company and total operation is a model of innovation and efficiency, and he has received lots of recognition and loads of awards for his achievements.

Ron has worked out innovative ways of employee compensation, psychological evaluating, material handling, and scads of other methods and systems relating to the workplace. At the same time, he has been very active in civic and community betterment affairs. He has been recognized for his achievements and garnered awards from the Chamber of Commerce, and from various local and state sources.

For a young man coming from the small community of Foster, Nebraska, and lacking a lot of more formal education, he has gone miles and miles and achieved great things in business and in manufacturing. In short order, Ron the student had become Ron the teacher.

He became very interested in material handling, storage, and work flow. He developed actual equipment, in addition to material holding racks, and cranes. He has worked out labor-saving systems utilizing the computer, minimizing scrap materials and fall-off, while saving valuable floor space. Always conscious of his fellow employees, Ron developed very ingenious compensation models and always maintaining high degrees of motivation. He worked out interview techniques that took into account differing personality types and factors. He has continued feeding his insatiable appetite for knowledge in many areas by investigating and experimenting at all times.

I remember old Gus Graske (G & G Manufacturing) telling me, "That Ron, he's a real 'comer!'" I wonder if Gus realized how "right on" that statement really was!

Horror Story No. 2

October 28 turned out to be a very unlucky date, in addition to being the date of Larry's fatal accident. Exactly one year later, in 1966, we experienced the worst shop accident in our history.

We were putting together a bridge for 72nd Street near Grover, and we had the main girders lined up on the floor of the north shop. There were about six of them, side by side, and about eighteen inches apart. Wilbur Herron and Duane Shupe were both between them, when someone, operating the overhead crane from the cab, accidentally hooked one of the girders with a chain grab under the top flange of one of the girders, spilling the girder over, and into the other beams.

Somehow, Duane, who is fortunately slender and on the small side, was able to get completely under the girder and was not very seriously injured. But Wilbur was pinned between the girders at the upper leg and was immediately screaming for us to get the beam off of him! It seemed to take an eternity to move the girder, but we freed him, and he was hospitalized in Blair for several days. For quite a while, doctors were not sure whether he would lose a leg, or perhaps both of them. He was laid up for a long time and was not walking for weeks.

But Wilbur recovered, and we were fortunate that we did not have someone killed. Believe me, changes in the way we handled beams were instituted, and eventually we were able to work these sections more from roller conveyors, meaning much less handling and therefore less risk to safety. We continued to operate with an undersized shop and without many tools and pieces of equipment that we could have utilized and that could have made us more efficient and safe.

This is about the time that we began to try to picture better ways of material handling, less hazardous, with hopefully less handling as well. It would be so much superior to think in terms of a fabrication line where the better part of the work could be done on top of a floor-mounted conveyor system. We had no plant engineering department and not even any specialists to help us. Whitney in Rockford, Illinois, was showing a beam line that incorporated just what we wanted—floor-mounted roller conveyors. This had to wait, since we had other problems to solve as well.

We were acutely aware of the limitations that our ground and buildings put on our business and on our dreams for growing the company. Our best chance for room lay directly to our west, and the property was owned by Ray Watson, who was formerly a part of Watson Bros. Transportation Co. They had the facility leased to Waste Management Co., and I didn't know the length of that lease. I finally hit upon a plan for an approach to Ray Watson that I thought might work. I knew that Ray was a crafty old curmudgeon and that he would immediately see that we were hemmed in and had no other way to grow, and so could really hold us up.

I made an appointment to see Ray, and when I went to see him, I noticed that he had lots of books piled up on a desk next to him, and all relating to properties that they owned around the country. From that, I knew that he was engaged in the systematic sale of all the properties that they owned, so I decided on a little "reverse English" to use on Ray.

I told him that we needed room and were looking at property out west. We had decided we needed to move. I suggested that he might like to buy our property, since that would give him a much larger area, and that it could combine with the old original terminal directly to the north, plus his own land to our west. As an afterthought, I suggested that he name a price for his, and that I would then name a price for ours, and we could then see if there was any room for a deal—if he didn't buy us, maybe we would consider buying him.

Ray said that he was in the process of selling off much of their holdings—terminals in Denver and other places—and that he had to go slow, since he did not want to incur too much tax liability in any one year. He would get back to me.

We didn't hear from him for weeks, and I was getting antsy but did not want to show him that I was too anxious to deal. I finally called him and tried to be "cool." I suggested again that he might want to buy our property. Again, he cited all these weighty problems with taxes and attorneys, etc., and he would get back to me. The waiting continued … and continued.

We stayed very patient, and it was a good thing. Finally, he indicated that he could talk again, and we had another meeting. At this time, he said that they had no interest in buying our property, but he would consider selling his.

Trying to keep from getting too anxious and too dry mouthed, I managed to ask him to name a price. I think he quoted me somewhere in the vicinity of $68,000, and I winced, acted shocked enough, and said we would call him back with an answer. I had to restrain myself from calling back immediately, upon my return to the office, but I waited, and I don't remember whether I dickered or not, but finally I told him we would pay him his price.

I know that I had feared he would really hold us up and ask about double that amount. He named a day for the transaction to take place, I believe in early 1968. I showed up without an attorney or real estate agent to assist me, and we did the deal in his office, with only his old office manager for a witness.

When it was all signed up, Ray looked at me, and with a crooked little smile on his face, said, "Williams, I think you just wanted to buy my piece all the time!" I just smiled and asked him if he was satisfied with the transaction. He said that he was, and I said that we were happy with the deal, too. But I do think that by using this approach, it made him tone down his asking price quite a bit. For certain, we didn't exactly have millions to spend.

Finally – Some Room to Grow

We could hardly wait to evict Waste Management, but we had to wait for a few months, as we had bought the property subject to the lease that they held. The only difference was, the lease payments came to us now.

We had built the craneway previous to this, so the building really amounted to framing in and covering about 200 feet of it for the structural addition. We could now plan for an extension of this crane to the west, over the newly acquired property. I guess we thought it was all the space we would ever need. Art Williams said we were "nuts," that we would "never use all that land!"

Now we could go ahead with our plans and not have to worry about the obligation that we had to keep the spur track to the west open for anyone else. I could see the plans Mike and I roughed out on the shelf paper those long years ago getting closer to taking shape.

We could build the south structural shop, and we employed Al Beyer to design and help us build the new shop building. Al had training as an architect but, for years, had worked as a steel detailer for Gate City Steel. He was a competent technician, if a little crazy at times, and tended to skim over some of the details. But he was a guy who believed in getting things done, and he was just what we needed at the time. We always worked with Beyer without a formal contract, just some rough estimates with the understanding that changes were inevitable, and we knew he was flexible and fair with us.

It never occurred to us to use an engineering firm to do any planning. We knew that we were capable enough at building things ourselves, and like an old country farmer, we just knew what we needed. Besides, if something needed some revising later on, we could just fix it on the spot.

We never checked the soil, although we knew there had been a creek through the area at one time. We erected our own steel. We hired K Electric to do the wiring. This was George Kuzela, who had formerly worked for Armature and Electric, and had worked on our "D59" building. We did most of the work ourselves, rightly or wrongly. We were lucky that we did not get an inordinate amount of settling from that soil. We made the mistake of not raising the floor enough to avoid flooding when the really heavy rains came. Many times we suffered hours of stoppage due to water covering those floors, which had to be swept up. But the net result was all positive, since we now had a lot more space for fabrication of sizable projects.

But, in one instance, we went a little too far. Back in 1959, when we had purchased the cranes from Cushman Motor Works in Lincoln, we had set aside the old antique model, which was a Whiting, and was a D.C. wired job. I decided to utilize this on the new craneway, and so we widened the bridge out from around 43 feet to 60 feet and asked Armature and Electric to rewire the crane and get it ready for new service.

This was a serious mistake, as the cost ran out of control, and, anyway, the crane was a real monstrosity. It had never been engineered to do what we were trying to do—the twin crane beams were way too close together for the new span of 60 feet, and the controls and powering, being DC was old and inefficient. We finally had to junk the whole thing, and I had to admit to a big mistake in attempting to utilize it. I guess the lesson was something like "don't throw good money after bad."

Another error was on poor Bill King, in that the main crane beams in the west extension were barely sufficient for the loads, and we got too much lateral deflection, which could have been disastrous. We came along later and beefed them up with bracing, which again was like the country farmer, making the changes "out of necessity."

Unfortunately, when we laid out the craneway to extend all the way to the southwest corner of our land, it caused us to end up encroaching on the U.P. right-of-way. We knew that the railroad would never object, or at least we counted on that, as long as we were active customers with freight moving in and sometimes out. In fact, we prevailed on the U.P. to build a bin wall on the south, to keep the bank from eroding and caving into our building. They built the wall, entirely at their expense. The railroad was counting on increased traffic to the site, which was only good business on their part.

But the 1960s was a decade of real growth for us, especially after 1965. It was a decade of turbulence, change, and, sadly, a lot of violence for the country. It was a decade of assassinations, drugs, war in Vietnam, and rebellion on our campuses and streets. Threats from foreign sources, Russia and Cuba, had to be dealt with, and a space race with the Russians was begun. Against the background of all this turmoil, with our kids growing yards and yards of hair, and "turning on and dropping out," it was surprising that we could keep our minds on the job at hand. There were times when we wondered where in the world our country was heading.

But, during the decade, we were able to hire some very valuable people, individuals who turned out to be great team members, and more than just valuable to the company.

In addition to Ron Parks, who we have already seen as invaluable to our efforts, we added Tom Barry in 1967, John DiBaise in 1969, and Lou Lesac in 1968. Tom came to us as a trainee detailer, but was later moved into sales. This was a fortuitous move and one that paid dividends for us as well as for Tom. John came from the same place as Ron Parks, Continental Can Co., and was recruited by Ron. He soon worked into purchasing, starting as an assistant to Mike Williams, but soon became the chief purchasing agent for us. John, through his loyalty and his diligence, has become a real star for the company. The same holds true for Lou Lesac, who has worked at various jobs over the years. He is now traffic manager and has done a great job organizing all of our shipping and receiving. These guys, and others, have been really team oriented and very appreciative of the profit-sharing program and its yearly results.

Al Ludwig Joins Up

A great addition to our company and its sales thrust was when we hired Al Ludwig. Al had been in sales for Wilson Concrete Co. for a number of years and had made lots of friends among the local contractors. He had a health setback which caused him some dissatisfaction with Wilson management, and he went to work for Ideal Cement Stone.

Mike and I knew Al from our association with the Field Club of Omaha, and we decided that he could be a very good addition for us. I picked Al up one afternoon, and we had coffee somewhere and then drove around town while we talked. The talking went on and on, and I guess I was going to keep him in the car until he agreed to come onboard!

It turned out to be a very good fit for Al and for us, and Al did a great job of keeping abreast of the market. He got his share and then some of the jobs that were out there for the bidding. He maintained all of those good friends, and the contacts really paid off in terms of increased business. He kept our estimators hopping, sometimes at too great a pace.

Al was highly thought of in the construction community, and he was able to get lots of valuable "feedback" from his many friends around town. His contacts were invaluable to us.

Finally, the time came when I thought the shop and the union would be "ready" to hear about a profit-sharing plan of its own. It was a gamble, since we still felt that we would have to battle over the terms, and we stated unequivocally that the union would have no say in the determination of benefits that would issue from the plan. There would be no disclosure to anyone of confidential internal financials, and no discussions about allocations to the plan. This was not a problem, since the union's sole objective was to keep the shop membership intact, and thus the dues stream intact.

After our negotiations, this was our announcement:

November 6, 1969

To Our Employees:

Today you received your first paycheck under the new two-year agreement reached recently by your committee and the company.

I think you are all happy about the wage increases, both now and the one that will take effect one year from now. It is more than four times greater than the largest single increase previously. The company is happy to grant these increases, as well as the fringe benefits, which include:

- Two extra half-day holidays
- Company paid insurance during an employee's disability
- Disability benefit (loss of time) increased to $60 per week

We think that as a company, we have an opportunity ahead to grow and prosper as never before. We are in an area which is growing rapidly, and where there is strong demand for our products. We have recently acquired land for expansion to our west. A major Omaha fabricator has recently withdrawn from the structural field, and another major Omaha fabricator chooses to bid only large jobs, many of which are miles from Omaha. This all points to more potential business for us.

We have plans for expansion. You are already aware of the craneway extension we are making to the west. We hope to start soon on a brand new structural fabrication shop, about 70 x 275′ in size. We also plan a new warehouse building, a new office building, and still another new craneway, probably located where the old east warehouse now stands. All of this is going to provide new job opportunities, and chances for advancement.

We have mentioned some of the provisions of the recent contract agreement. We have saved what we feel is the most important feature until last, and this is the Profit Sharing Plan for Hourly Employees. In fact, this will be the most important single benefit ever to be added to any of our past agreements.

This plan has been designed so that every employee will have the opportunity to share in the company's future, but it all hinges on profits.

It goes without saying that when we eliminate waste of time, waste of materials, and when

> we work safely, we are more efficient—and therefore more profitable. The big difference now is that you will have a real stake in these added profits which are created.
>
> It is not our purpose now to try to fully explain here all the provisions of our Profit Sharing Plan. This will be done later at an appropriate time.
>
> We have had a big improvement in employee attitude, in morale, and in general cooperation over the past several years. We want this trend to continue, and to grow—and for all of you to be proud of the place where you work. We need and want good, faithful employees who will stick with us, for our future is only as good as our skills and our efforts.
>
> Hugh W. Williams
> President

Attitudes and morale were getting a little better, but still there was too much of a union mindset among the men. They were not used to anything else, and they were going to be from Missouri—Show Me!

Then, in 1970, we were able to add Gordon Cizek to our growing family. Gordon had already been in the structural fabrication business with Gate City Steel for some twenty-two years, so he was a real veteran. Rather than coming to us as somewhat worn out after that time, Gordie was very enthused, happy to have this new connection, and raring to go. Gordon had a very sound background in estimating and was very methodical. He brought us systems that helped us get away from "pound" pricing and into a better way using shop manhours. He is like an old war-horse and still puts in quite a few hours at the time of this writing, despite being officially "retired." Gordie knew from his days in the shops at Gate City, and long experience with the old hands there, how to put hours on an estimate.

Gordon has truly become a legend at D-W-S and is still active after more than fifty-eight years in the business, counting his service at Gate City. He is apparently the counterpart of our iron-man Bill Fuhrman in the shops. Gordie is still an active roller skater, pilot, and rides his own motorcycle! He has been a tremendous asset for our company over all these years.

For a time, we called the combination of Tom Barry and John DiBaise our "mafia sales team," and John, perhaps responding to the humor of all that, sort of became our office joker. He usually was good for a joke or two, and usually pretty ribald. Later, John became a fine "disc jockey" at some of our better, evening Christmas parties, or maybe running a "bingo" competition and entertaining the whole bunch of us.

For his part, Tom was the "Beau Brummel" of the office staff. He took a fair amount of kidding for his sartorial splendor, but it paid off when he was out "on the street" and calling on customers. Tom and John, both tremendous assets to the company, and helping us grow.

After George Hoody, who had retired in 1958, we had a succession of characters try to become shop superintendent, or just plain foreman. We mentioned Leroy Shoell, and his acceptance of sloppy workmanship and lax methods, leading to his firing in the early 1960s. I brought down a guy named Al Riskowski, who had been a foreman and parts manager for Trailmobile. We gave him a try, but Al just did not have the background or the skills for this kind of a job. He tried hard, but we finally had to let him go, too. Then we discovered the diamond in the rough that we had all along. Sometime, before 1965, we awarded the job to Duane Shupe.

Shupey

Duane Shupe had been with us since 1955—the only employee left at that time who had more seniority than I had. Duane started in the shop as a burner. In those early days, we had no plate shear over a quarter inch, and little in the way of sawing sections, either. So the plate and miscellaneous burner was a busy, busy man.

I will never forget Duane, who had to squat like a baseball catcher most of the day, setting up and adjusting his "bug" that ran on the tracks, which was our burning assembly. How he could maintain that position all day, day after day, was a marvel. He put out an unbelievable quantity of work and never complained, never missed a day of work. No one was more cheerful, no one more willing, and would work all kinds of hours if needed. Duane came originally from a farm background, up around Sioux City, and he understood what a day's work was—and then some!

I remember when we first projected Duane in the job of superintendent. We knew that he was not a layout man, but neither had the last two men owned that skill. We had Bill King, who could make patterns when we needed them. But we worried whether Duane's co-workers would take their orders from him, since he had worked so long alongside them. Would he have their respect and could he make demands of them? We decided that he had a good chance, and we knew he would give it his best shot. We were right—or I should say, more than right.

No one in the company has contributed more in their employment to the benefit and betterment of D-W-S than this man. He never changed in all those years, always positive, and almost always bringing the job in on time, and as promised. Only one other man, Wilbur Herron, who was a mechanic-lead man, could put out the amazing quantity of work that could compare with Duane's output.

It wasn't long before Duane was laying out plate work as well. He picked up the skill from textbooks that he found and taught himself—on the job, self training! He became pretty adept at it, too, but the quantity of that kind of work was becoming less, as we managed to get more and more structural work. So he quite filled all the requirements of the job and managed to get the respect of the men under him, too.

Except for one, however, who did not carry the requisite respect that Duane was really entitled to. On this occasion, we had a man who had broken several of our rules of employment, was a burly kind of guy, and had become a real trouble maker in the shop. It came to the point that Duane knew he had to fire him, and I don't think he looked forward to this. Sure enough, when he fired him, the guy took a big swing at Duane and knocked him right to the deck. It was a sucker punch, and the guy just marched out of the door and kept right on going. Duane was pretty embarrassed about the whole thing, but he said it was really not all that unexpected.

Banking on the Best Bank

There was turmoil of a sort on the local scene as well. Our banking source, which had always been the First National of Omaha, was undergoing turmoil as two families vied for control of the institution. It resulted in a difficult situation for us, in that we could not get the kind of support for financing at competitive rates, to the point where we were forced to make a banking change.

Jack Shonsey had an interest in the American National Bank, and he convinced us that he could take good care of us and make us loans at competitive rates, and we made the move to his bank. This was a big help at the time, but as time went on, we found that they were not able to extend us a large enough line of credit and had to use a correspondent senior bank, the Omaha National, in order to service our needs.

So, we had left the First National Bank, with which we had been for possibly eighty or maybe ninety years! This remained our banking affiliation until 1980, when we were welcomed back to the First, with a very advantageous set of terms. They have been our good partners ever since, and always a huge help. I remember that Mike and I were able to negotiate the terms without having to sign personal guarantees. Our CPA and my close friend, Don Darst, always marveled at that and said he had not a single other client in similar circumstances that got by without those guarantees. The man who called on us, and recruited us back to the First, was Chuck Fries, a very astute guy and a terrific banker.

The First stuck by us during some very tough years, in 1982 and 1983, when we were losing money for the first time in our history. They did ask for personal guarantees, but returned us to our former status a short time later, without the guarantees. Of course, we had little idea of how that bank would grow and how large it would become.

The First had always been the number two or three sized bank in Omaha, following the old Omaha National, and the U.S. National. The Omaha merged with the Lincoln bank and the U.S. sold to the Northwest system of Minneapolis. Nevertheless, the First grew so rapidly that it overtook all of them and has now become the premier bank in town, as everyone knows.

This improved banking affiliation was the tonic we needed, because we were starting to bid larger and larger projects, and these required more working capital. We did not grow so fast, apparently, that the bank ever balked or expressed any real doubts about us. They took care of our needs, and we were actually saving money on our interest expense.

During the high, high interest rates of the late 1970s and continuing into 1980, we had our difficulties with the trade. Certain accounts liked to use our money, and the rates of interest were a strong incentive in this direction. We were having the most trouble with the particularly large firms—those that had the most "clout."

We learned that we had to be very firm and had to enforce our terms, which called for interest to accrue on overdue items. We were at a disadvantage, because we did not have a strong credit manager. We knew that we needed to have someone who could really get tough over collections, and we did not have that man. No one likes this task and this responsibility. We did the best we could. We were really happy when rates began trending down, and the heat was reduced.

It was May 9, 1980, and we were welcomed back to the First National Bank. They would make available a $1million line of credit at 1 percent over the FNB's prime rate. This was to be a revolving line of credit, with take downs and payoffs each day. They had made a study of our borrowings over the past year, and they estimated that this new scheme would reduce our average borrowings for the year by between $50,000 and $100,000. They estimated that based on our prior year's borrowings, we would save about $13,500. It was a good proposal and was far less unwieldy for us also.

We also struggled with retainage on large contract accounts. We tried manfully to write out any of these retainage provisions and were finding some success. We worked with Kiewit and, for awhile, had these eliminated. We found that some of our customers abused these to extremes and that it was always so during periods of high interest rates.

We were doing a lot of business with Kiewit, and sometimes it was a good learning experience for us. There was a time when we were not in tune with scheduling. We did not have a coherent system of manhour accounting, which left us at a definite disadvantage when trying to project and to schedule work into the future.

Kiewit, at the time, was adamant on the subject, and we had no choice but to comply. It was a little painful, but in the long run, we were forced to think in these terms, and we learned a lot about our own operations as a result. Things went much, much smoother. Then, we began to ask, and even demand, this kind of scheduling on the part of our other contractor accounts. It was easy to smoke out the ones that were not well organized. They just were not the top companies or operators in the field.

Doing business with Kiewit was just a little bit more demanding at this time. Once you were the apparent low bidder on the steel on a certain job, you could expect to be "hot boxed" up in the office of the Omaha project manager, who at that time was Bob Tomek. He was not satisfied to talk to your salesman, even if he was the sales manager. They wanted face-to-face assurances on delivery, and they wanted to perhaps find a few dollars of "savings" before inking the contract.

One day, I happened to see an ad of some kind in one of our trade magazines that was around the office, and it contained a picture of some guy in a dungeon-like situation, and I clipped it and gave it to Al Ludwig. I scrawled on the ad something like, "How it feels to be cornered in Tomek's office." I didn't expect Al to give that to Tomek, but he did, and Bob really had a great laugh about that. After that, the atmosphere in his office was quite a bit more relaxed and low-pressure.

I can remember more than once making iron-clad promises of delivery in that office and then later, in private with Gordon Cizek, saying, "We can do that, don't you think, Gord?"

Finally – Real Bricks and Mortar

Between the time we bought the Watson land from Ray Watson in 1969, until 1976, we built the first section of the South structural shop, then added onto that shop, then completed the crane yard to the west, then built the West warehouse building, and then the present office building fronting on Hickory Street. Also in that period, we tore down the old wooden East shop building and replaced it with a craneway from Hickory Street back to the UPRR spur track.

Most of this expansion and new construction we did out of cash flow. Some was financed using funds provided by Hickory Partnership, which was the partnership between Mike and myself. We owned the office and the West warehouse and leased both of them to the company until the company bought them from us in 1996.

Most of this work was designed by Al Beyer, our friend and architect. We subbed out the plumbing and heating contracts, the electrical contracts, and, in the case of the warehouse, the steel erection and sheeting. At times, this was pretty exciting work trying to find old, buried sewer lines, telephone cables, and gas lines.

Digging for the footings for the West warehouse brought quite a response from the phone company and from Strategic Air Command (SAC) headquarters. It seems that we were within inches of the buried cable, which was the primary line for the SAC red phone. No emergency alarm went out to warn that the Russians were coming.

We were very busy at the time of the construction of the West warehouse, so we opted to use American Buildings components. I have forgotten the name of the contractor we hired for the erection and sheeting of the building, but about halfway through the job, he went broke, and

Overview of buildings and yard, 23rd Street plant about 1975.

we were left to find a way to finish the job. We ended up putting most of his men on our payroll and financing this project ourselves.

The interesting part was that as we proceeded along, we would lose a man out of this crew, and this happened about four times. Each time, we noticed that production did not suffer a bit, and actually seemed to get a little better. So the long and the short of it was, we ended up with a finished product for less money than we had anticipated.

During the 1970s, business was quite good, and we were trying to step up to do larger and larger jobs. Finally, we landed what was our first-ever thousand-ton job—the Creighton Dental Building. We were elated, and I had a desk sign made for Mike, who would be doing the steel buying. It said, "I'm tougher than the Israeli Air Force," a reference to the three-day war when Israel whipped Egypt's butt so soundly. This

Present main office of Drake-Williams Steel, 23rd and Hickory streets.

was a Kiewit job, and although we made a few mistakes, it turned into a great project.

This was a period of time in which a lot of attention and concern was directed to the problem of imported steel. We were buying strictly from domestic sources, however. One day, I saw a Bethlehem ad in *Iron Age* magazine, which was in the form of a letter from Lewis Foy, Bethlehem's chairman. He took it upon himself to warn his customers not to buy foreign, and that if they did, Bethlehem would have a long memory, with the threat being that at some future time, when steel became scarce possibly, these customers would not be likely to get any steel, or maybe minimal consideration, from Bethlehem.

I was furious at the presumption, and we were already tired of the major mills' tactics when it came to knuckling down to their unions, and then raising prices to the trade to pay for their weaknesses.

I sat down and wrote a scathing letter directed to Foy, the first I had ever written to any top mill official. We suggested that instead of sending us a lot of bad advice, he might want to finally stand up to the steel unions and draw a line in the sand once and for all.

We received no answer from this coward, but instead, he sent the sales manager from their St. Louis office all the way to Omaha to see us, and his story to try to justify Mr. Foy's words was totally lacking in common sense or intelligence. Mr. Foy did not serve as chairman for too long after this.

Running Out of Room

In 1981, an opportunity to the west presented itself. Always needing more space and more facilities, we managed to purchase the south end of the Omaha Steel Castings (formerly Omaha Steel Works) property at 48th and Leavenworth streets. This plant had been the main fabricating facility for Omaha Steel but had been idle for quite some time. Naturally, it was not in very good repair, and we spent quite a bit on electrical work, miscellaneous repairs, and updating of utilities. One of the reasons that we wanted the facility was the spur track off of the Missouri Pacific Railroad. Soon, it became the property of the UPRR by acquisition from the MoPac.

We had ignored A.I.S.C., the American Institute of Steel Construction, for many years. We thought, for one thing, the dues were prohibitive and the benefits not that apparent. Finally, we decided that we should not be going along for a "free ride," since that organization certainly promoted the interests of the industry in general, so we joined the group. We organized a committee to prepare for a certification

American Institute of Steel Construction, Inc.

This is to certify that

Drake-Williams Steel Inc.
Omaha, Nebraska

has the personnel, organization, experience, procedures, knowledge, equipment, capability and commitment to produce fabricated structural steel of the required quality for

Conventional Steel Structures
Simple Steel Bridges
Complex Steel Building Structures
Major Steel Bridges
Sophisticated Paint Endorsement and
Fracture Critical Endorsement

President, AISC

Auditor

Initial Certification: July 24, 1990
Audit Date: June 4, 1996

Valid Until: July 24, 1997

A.I.S.C. certification earned in 1990 and continuing to this day.

examination, which was a big undertaking. We were accepted on the first exam, which, as we learned later, was quite a rarity.

Because of the adverse business conditions that followed in the middle 1980s, we did not fully utilize the facility for some time. For a while, we planned on making it our bridge fabricating facility, but this was not to be. We probably could have stayed with it, and persevered, but we had a hard time booking enough bridges, and the prices were not an advantage either.

Then, eventually, after the Union Pacific (UP) take-over of the belt line, an announcement was made that changed everything. The UP

Operations at 48th and Leavenworth, which D-W-S owned from 1981 until about 1994, formerly Omaha Steel Works.

served notice that they wanted to abandon the entire belt line route. Traffic for them was slim, and there were fewer and fewer customers along the route. Lyman-Richey Sand and Gravel was one, but even that going concern was using less and less rail.

We tried to protest the abandonment, but we could see that the UP was far too powerful to oppose, and it was an inevitable fact of life. After this, we tried to maintain that the land underlying the rail line and the rail spur should revert to the land owner alongside the line. We finally took the UP to court over this, and there was a trial on the matter.

We presented our case, and much supporting evidence from parallel cases, and we felt that we had presented a much better case than the UP attorney, who was Brenda Council at the time. It was not a big deal in terms of value to us, but for the UP it was much larger. If we had prevailed in this, the UP would systematically lose all the ground over the entire circuit, and it would doubtless be cited as a case precedent in other abandonment proceedings, everywhere along the vast UPRR system.

Although we had made indisputably the better case, the Union Pacific prevailed, and the land remained in their possession. We considered an appeal to the Supreme Court, but decided against it. So we dropped the matter and began to weigh our options on the matter of even remaining in the West facility.

We had completed the purchase of the plant, which was $500,000, and we decided to put our rebar operation, which was doing well and growing pretty fast under the able direction of Charlie Wheeler, into the north end of the plant. The offices in the middle building at the south end were a fit for the staff of managers and detailers. We had invested in an upcut line and conveyors for the rebar operation, and we moved all of this to the West plant. This was not a bad solution except that we no longer had rail service, and that was a big drawback.

During those rough years in the early 1980s, we resorted to different things in our attempts to generate business. About this time, the city of Omaha had decided to build some bus shelters for the convenience of riders and to help in encouraging people to come back downtown for their shopping. The projects were advertised in the Dodge reports, and there were going to be six or eight of them built.

I knew that our estimators would not look twice at these jobs, because they were really not our cup of tea. They were more in the category of ornamental, or architectural work. We had precious little on our boards for new work, and the work that was coming out was going absolutely dirt cheap. I sent for the plans and told the estimators to take a shot at it. The thing was, these were not bid in the usual way, but had to be the finished product for the contractor to pick off your truck and set down on some anchor bolts.

This meant we had to figure on furnishing all the various components such as the heavy brass handrail around the inside, the necessary electrical wiring inside, and the copper roofing. All these things were expensive items, and to top it off, the painting of the steel materials had to be extra special, sandblasted, and close to perfect.

So we put all the pieces together, and as luck would have it, we turned out to be the low bidder. They turned out to be pretty profitable for us, especially compared to some of our other jobs at the time.

But the biggest surprise of all to me was the quality of the work that we were able to provide, and the finished appearance—a degree of finished work that I did not think we would be capable of. I gave the shop a written commendation for their efforts and gave a sigh of relief that we could successfully pull off those jobs.

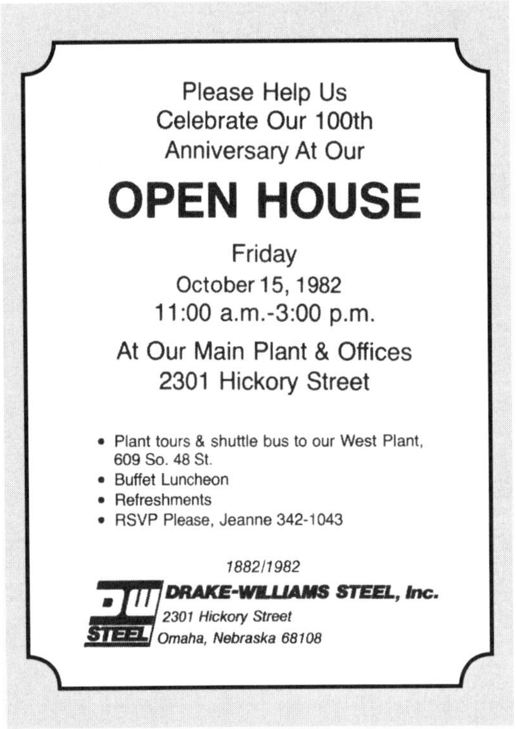

Please Help Us
Celebrate Our 100th
Anniversary At Our

OPEN HOUSE

Friday
October 15, 1982
11:00 a.m.-3:00 p.m.

At Our Main Plant & Offices
2301 Hickory Street

- Plant tours & shuttle bus to our West Plant, 609 So. 48 St.
- Buffet Luncheon
- Refreshments
- RSVP Please, Jeanne 342-1043

1882/1982
DRAKE-WILLIAMS STEEL, Inc.
2301 Hickory Street
Omaha, Nebraska 68108

The invitation to our 100-year open house and celebration in 1982.

A "Flat Tire"

Another diversionary effort that we made about this same time did not turn out nearly as well. This evolved as a result of David King (Bill King's son) delving into computers and computerized detailing. David had worked for us as a detailer and later had been put in charge of our computer operations.

Dave found out about a promising program that he checked out in Philadelphia. He brought back the story, which had impressed him and which he felt was something we could utilize, even though some of the program, called "AdSteel," was still under development and was not complete.

We purchased a sizable IBM mainframe computer especially to run this software, and we were involved to a greater extent financially than I was comfortable with, but we were putting a lot of reliance in young King. He was a very bright guy, and he said that he was confident that the system would work, was very good, and could be utilized to our great advantage. Not that he didn't think that his efforts plus our investment would not pay off, but we did not know really where all of this was leading.

We were to find out soon quite a bit about the limitations and shortcomings of the software. The system could detail simple beams and columns, but did not do the difficult things—the things our detailers spent the most time doing, in other words.

Again, with regular work becoming more and more scarce, we figured that if this software could really do the job, why not use it at a pace more up to its capabilities and volume and offer a detailing service to the trade? We should have known better, since we had not given the systems a thorough enough testing on our own work. Dave King was confident of the system, so we decided to go ahead.

We actually started a separate company and then began hiring extra detailers. We hired a young whipper snapper to sell the service to the other fabricators, and he took on some jobs. I don't remember what we called the company, but almost at once, we had a disaster on our hands.

The drawings were not complete, and we were not able to check out just where the system was dropping, or missing, vital pieces. We lost a fair amount of money, all in all, and then we had to face up to the fact that we had an unfinished and untried bunch of software, and a huge mainframe computer that was not at all what we needed for the rest of our operations.

It was a tough thing to have to face up to, the cost of the computer being huge, and the rest of the expense a total loss. David King left soon after this, taking a job with OPPD. But we learned a lesson in the process. What was difficult to take was knowing that we had senior and seasoned management that should have seen the pitfalls before it was so late in the game. Never again, we said! All this happened about the time that minicomputers and PCs were displacing many of the huge mainframes in offices everywhere.

It was a bitter pill, but as president, I had to accept responsibility.

Opportunity Comes Calling Again

Another problem we were having at the West plant was with our neighbor to the north—the outfit that had sold us the property in the first place. Little annoying matters kept creeping up, nothing important, but just aggravations. They seemed to feel that they should have access across our property at any time and access to the gasoline pumps also. This was bad for security reasons during off hours, but should not have been such a problem.

Then, about this time, I ran into Monte Matz, a realtor in town, and he asked me if I would like to hear about the property on North 11th Street, formerly Gate City Steel.

I at first did not take this too seriously, primarily because I figured the facility would be way out of our price range. Matz's company had the listing. The asking price was about $1.3 million. He had a brochure with a partial description of the layout, and so I did a little daydreaming.

I talked to Mike about it, and he felt about the same way I did—that it was pretty pricey for our use. I mentioned it also to my two sons, Dave and John. At first, they were not really enthused. But the idea began to eat at me, and I talked to Monte a couple of more times about acquiring the site.

I laid out a rough plan for how we would utilize the property if it were ours. My plan was long range, though sketchy, and did not contain any hard and fast numbers. I contemplated buying a lot of equipment, including a drill line that was just a sort of "pie in the sky" item for us at that time. Other equipment such as a shot blaster was another item. Money would be needed for insulation of the main east-west bay, and we would also need a heating system. I did not include the needs for updated and heavier electrical service, for compressed air, natural gas piping, roofing, and much more. I had no idea of the total deficiencies we might find.

Finally, all of the Williamses at last seemed to be in favor, if tentatively, and

Incentives Keep Firm In Omaha

BY JIM RASMUSSEN
WORLD-HERALD STAFF WRITER

Nebraska state tax incentives convinced an Omaha steel company to expand within the city instead of accepting offers to move to Iowa, company officials said Friday.

Drake-Williams Steel Inc. said it has purchased the former Gate City Steel Works at 1601 N. 11th St. The company said will expand its structural steel fabrication operations in the former Gate City plant.

Drake-Williams plans to hire 25 to 30 people within the next year, said David Williams, company vice president of operations. The company now employs about 100 in Omaha.

Tax incentives contained in Nebraska's Investment and Job Growth Act, passed in 1987 as Legislative Bill 775, kept Drake-Williams in Omaha, said John Williams, the company's vice president of administration and David Williams' brother.

The law offers tax credits and refunds to companies that meet levels of capital investment and job creation in Nebraska.

"We had attractive and tempting offers to move our operations to several Iowa points," he said. "Without the benefits of LB 775, we would have moved."

Terms of the plant sale were not announced. Drake-Williams bought the plant from Gate City, which still has operations in Omaha. Gate City had not used the plant for several years. The plant, built in the early 1930s, covers 185,000 square feet and sits on 13 acres.

In applying to the state for LB 775 benefits, Drake-Williams said it planned to invest $3.7 million and add 30 jobs at the plant.

Drake-Williams will continue operations at its main plant, 2301 Hickory St., as well as expand into the former Gate City plant. A third company plant near 48th and Leavenworth Streets may be sold, David Williams said.

World-Herald *announcement of purchase of former Gate City Steel plant on 11th Street.*

Aerial view of new North plant at 11th and Seward, 2004.

we began negotiations through Monte with Progress West Real Estate. Gordon Cizek, who had spent about twenty-two years at Gate City before he came with us in 1970, let me know that he was totally in favor of the move, if we could swing it, and said it would fit our needs like the proverbial glove.

Several things worked extremely well in our favor. At this time, real estate of this kind such as warehouses and large industrial structures were a big glut on the market. Property was not in demand and was just not moving. The real estate legislation in 1986 had taken a lot of opportunity for depreciation advantages out of everyone's grasp.

Another problem was the concern about asbestos in old buildings and the cost of removal and remediation. Underground leaching of chemicals, gasoline and oil, and other elements were a pitfall. These things all helped us work on the selling price and the terms.

Negotiations stretched on and on, and 1991 turned into 1992. An admission must be made that it brought us quite a bit of satisfaction to be more or less dictating the terms of the purchase and the terms through an agreement written by our attorney in the matter, Sal Carta. The price was negotiated down from the asking price of about $1.3 million to the final one of $625,000.

Perhaps, according to Dave Williams, if we had known the extent and cost of all the necessary repairs, modifications, and improvements to the facility, we might have taken a different view, or possibly not have bought it at all. My brother Mike and I have always considered it a great move, and one sorely needed by the company to take up the slack of companies quitting business, such as Gate City, Northwestern Steel, and Omaha Steel Works. At any rate, with all of the changes, additions, and improvements, it is now a first-class facility and truly a model fabricating plant.

Very early in January 1993, I wrote the following letter, on behalf of my brother and myself:

> To All Fellow Employees:
>
> January 5, 1993
>
> After over twenty-seven years as President, and thirty-six years service in total, I have resigned my office and all active participation in the company.
>
> My brother, Mike, tenders his resignation also, after serving for similar lengths of time.
>
> Much was accomplished during this period, and healthy growth has resulted. We started as No, 3 in the state, and have now become probably No. 1.

We have had the good fortune to have extremely good people who have worked for DWS over these years. These are the results and accomplishments of the work of all you fine people. We are grateful to have been associated with you.

Now, the new team officially takes over. Dave Williams has been elected President and Chief Financial Officer; John Williams is Vice-President and Chief Executive Officer.

Much remains, and the future is bright. We are in an enviable position, and it is exciting looking ahead. We have a great group of employees and a real winning team. We now have the physical space and equipment. Mike and I remain heading up the Board of Directors to help us achieve the next goals and assist in any way we can.

Yours very sincerely,

Hugh W. Williams
Michael C. Williams

FOUR

FOURTH GENERATION: NEW ERA, NEW OPPORTUNITY

Explosive Growth and More Change

THE FOLLOWING DESCRIPTION WAS PART OF THE INFORMATION handed out at our Open House that was held in 2002. It is indicative of the vast growth and expansion that has taken place in our company in the last approximately ten years.

General Information

The Company operates out of two locations, The North Plant located at 1602 North 11th Street and the South Plant at 2301 Hickory Street.

The DWS North Plant Facility was purchased from Gate City Steel in 1992.

The North Plant property consists of 13 acres of land, 218,000 square feet of shop space under roof, 61,000 square feet of storage area under roof and 8,600 square feet of office buildings.

From 1993 to 2002 a number of major plant improvement projects were completed and equipment purchased and installed.

The North Plant houses the company's Structural Fabrication Operations, Contract Sales and Project management, Service Center sales and warehousing, Purchasing, Reinforcing steel shop, and the Engineering and Detailing office. The north plant works on a two-shift operation.

The South Plant houses the Corporate Offices, MIS and Accounting department. Rebar sales, Engineering, detailing and administration. Shop operations at the South Plant involve Rebar city sales, Structural rolling equipment, storage and warehousing. The South shop operates one shift.

Staffing: Total Employees at both locations is 174.
North Plant Shop has 66 employees on days and 25 on nights.
Rebar has 13 shop and 13 office personnel.
QA has a staff of 3, while Engineering has 14.
The Office Support Staff has 40 Employees.
Safety is our highest priority at DWS.

> In 2001 DWS earned the Safety & Health Council's Award of Honor with Distinction for 5 consecutive years of "outstanding occupational safety & health program results and achievements." Our safety program consistently ranks in the top 5 in our Association memberships—AISC, CRSI, and CFA. Shop Employees have weekly safety meetings and our Safety Committee meets monthly.
>
> DWS fabricates and processes over 31,000 Tons per year. 14,500 tons of structural steel, 12,500 tons of reinforcing steel, 4,000 Tons of Service centers steel each year. DWS stocks and keeps in reserve 4,500 Tons of Steel. Over 60% of this is Beams.

What followed at this point was a comprehensive list of plant equipment, including cranes and rolling stock. The list was extensive and was an indication of the heavy investment in plant and equipment made since the mid-1990s. It represents a huge investment in D-W-S's future, and the future of all of its employees. The Open House was a big success and well attended. I'm sure it was an eye opener for many of our friends and customers.

New equipment in the form of automatic beam drill lines and punch lines, along with the large structural saws were especially popular to the visitors and guests for viewing. D-W-S staff members showed obvious great pleasure in "showing off" their facilities and capabilities to customers and friends alike.

When my brother, Mike, and I retired in 1993, we had become a little tired, and therefore somewhat satisfied with the status quo. Human nature being what it is, this was the situation. Behind us was the next generation in Dave and John, and they were growing anxious to see some changes, and some new life pumped into the organization.

They could feel the wheels grinding to a crawl, and they knew that others in the organization were not happy with just maintaining—not growing. There is an old saw that if you are not going forward, you are really going backward—that there is no such thing as standing still.

Long before we actually formally retired, they made it clear that time was wasting! Both Dave and John had served in many different capacities and done many jobs—billing, costing, scheduling, estimating, just to mention a few. We knew that they were prepared, and they had a good grasp of the business. Besides, they had good educations before their working careers began.

The extensive amount of equipment and facilities added is mostly a tribute to Dave and John's initiatives and foresight. The rework on the North plant facility was huge and extensive, since it had been allowed to run down for many years. The expense was great and way more than we had really been prepared for, but still the finished product was well worth it. We now had the luxury of more space than we had ever known and dreamed of. This was not just an expansion, it was a quantum leap for the company.

These are the physical, or more apparent, changes and improvements in D-W-S. Much more (and more important, really) have been the changes and improvements in the "inside," the organizational and personnel (human resources) of the company. Systems, such as performance reviews, job scheduling and costing, and many other areas important to the efficiency and progress have been initiated and have been vitally necessary to promote orderly growth.

These changes and improvements, all brought on by Dave and John Williams, have been like turning an old tin lizzie into a true "muscle" car. Our sales were not merely growing, they were truly skyrocketing upward. I look at it in real amazement and just wonder where they learned all these methods and techniques.

I remember something my father said to me years ago, when I was a brand new employee. Someone in the office offered that I was really bringing in the sales and building a fire under the company, and he said, "A new broom sweeps clean!" I guess that is true once again.

It didn't take long for the company to put all this new equipment and facility to good use. But first, the old east-west craneway had to be worked over. Settlement had always taken place beneath the main A-frames, and therefore keeping the crane rails level was impossible. This also made the system unsafe and, of course, inefficient. Some of the actual bridge cranes had to be renovated and others discarded permanently and replaced.

Engineers and contractors were hired to put this facility back in good shape. I remember our Larry McMahon talking about the problems they had years before with that crane system. In the days that he was at Gate City, he told about how they just kept driving shims under the base plates of the A-frames to try to level the rails. Soon, however, he said that more settlement would take them down again, and the process would begin once more. He knew that the plant had been built right on top of what had once been a small lake on the property.

Finally, after extensive and expensive remedial work to the footings, the settling was stopped, and the craneway was made safe and sound. The chronic problem that Larry McMahon had said was impossible to stop, the continued settling, had been solved.

In addition to these required fixes, the electrical in the entire plant needed enlarging and upgrading. A new and larger air compressor was purchased to pipe air around the plant, and piping had to be replaced in many instances.

A new radiant heating system was installed, and roof insulation was installed. All of this and more was necessary due to the rundown condition of the facility. New equipment was purchased as well. I researched shot blast equipment, which was purchased and put in place. Conveyors were added, and much, much more.

The rebar downcut line had been moved from the West plant at 48th and Leavenworth into its new home, where it occupied its own separate shop within the plant. Other shears and benders were added, and under the very able direction of Charlie Wheeler, the reinforcing division took off on a new spurt of growth. Rebar took great pleasure in setting records virtually every quarter, and they adopted for their new motto: "Pour it on us!"

Sales and tonnage records were being set and broken again in other areas as well. Tom Barry and his department were having lots of success in sales to railroads. The Union Pacific soon became one of our best customers, and we broke into another important account in the BN, Burlington Northern.

Of course, structural sales were humming as well, and the company was now in a position to take on almost any job. When the First National Bank announced they were going ahead with a forty-plus-story tower downtown, a strategic decision was made to try to ensure keeping the job in Omaha, and to prevent an outside fabricator from stepping into the competition. The decision was to share the job with our main competitor, Paxton-Vierling Steel. PVS agreed with us, so a deal was made for them to do the building columns and D-W-S to do the beams. Now, both firms can claim they "did" the job—and it is an impressive one.

In recent years, D-W-S has lived up to one of our stated missions, which is to be "the fabricator of choice" for the local contractors. Davis Steel Erectors has acknowledged as much to our company, and others, including Peter Kiewit & Sons, have seconded that sentiment as well. The company has remained aggressive, opening a sales office in Des Moines and, lately, one in Denver. John Williams has moved there, as have other support personnel to emphasize our dedication to future business in the area.

When You're Old, You're Old

Here is a "note" that I sent the company, responding to the newsletters that I have enjoyed, and that keep me up on the doings and advances being made right along. It was included in our newsletter, as follows:

"Now ... and Then"

When you're old, you are old, and making comparisons just comes naturally all the time, which is what I'm engaged in at this moment. I have remarked before about the amount of pleasure I get from reading every "Newsletter" that comes my way. To see the company grow and prosper is exciting and deeply satisfying to this old man. When I see many of the people that I hired doing so well, and excelling at whatever they are doing, it is a real "trip." The physical growth in plant and equipment is exciting, of course, but to see the people developing and maturing is even a bigger "payoff."

I wish I could take each one of you on a journey back in time, back to when my brother and I made the decision that we had to change, and to find something we could sink our teeth into. We were going nowhere fast in the late Fifties, trying to build a business on a vanishing industry that was reliant upon steam power. We made pressure vessels, repaired steam boilers, and built breechings and smokestacks. But, the end of these businesses was surely in sight!

However, in the late 1950s, Omaha and the region began to get some industrial growth—the first we had seen in a long, long time. Western Electric (AT&T) built the facility in western Omaha, which was huge for our town. We decided we needed to get a piece of the construction market, and so, we would become structural fabricators. Frankly, we did not know really how to begin. We did not know how to bid a job, much less how to do one. That is when I hired Bill King, who was a civil engineer working at Wilson Concrete Co. Bill had had limited experience in Wichita with a small fabricator before he went with Wilson. At least it was something on which to build.

Ernie Vrana was a friend of mine, and he gave us our first steel job, which was for McKenzie Pontiac, south of us on 24th St. (Not the big showroom building, but the smaller one north of there.) We did not know enough to submit shop drawings for approval, but somehow we got through the job, and, while still learning, began to bid others. The first few jobs were a nightmare, as we had guys who had been boilermakers and sheet and plate fabricators now trying to read and interpret shop drawings. We made columns backwards, we made them a foot long or short, we made beams upside down, we mismarked items, we forgot the field bolts … in short, we made every mistake that you can imagine. But, we somehow got by, and slowly … very slowly … we learned. Ernie died a couple of years ago, but his son and son-in-law are very capably carrying on the business. Oh, did I mention? The reason I knew Ernie was that he dated my sister, Janet, when they were both attending Central High School!

Our facilities? We had next to none. We had not a single overhead crane. Where the east craneyard sits was an old, wooden shop with dirt floor. We had a Cincinnati quarter-inch shear, a Dreis and Krump quarter-inch leaf brake, and a couple of single-end Cleveland punches. We had some decent (if old and slow) plate rolls, and an old International straight truck. We did not own the west two-thirds of the property on Hickory Street, the part where the present office now sits. We had very little room (about an acre and one-half) and next to no equipment. Believe me, Gate City, Omaha Steel Works, and Paxton & Vierling Steel were not losing any sleep thinking about us. So, what was our advantage over these giants in the business? That is the point of this tale, and actually the moral of the story.

Of course, the answer was P-E-O-P-L-E. Not extraordinary people, no geniuses, and certainly no big money! We had people who simply persevered with very little and who were loyal to their company and to each other. We didn't think about the odds against us, or the possibility of failure. Failure was NOT an option, and every little battle that we won was savored like a BIG success! So the moral? A company IS its people—no more, no less. Not bricks, not steel, not machinery, but PEOPLE.

So, that's my little story, but there is still more to be learned. That little start-up, tiny competitor company that may be showing up on your radar screen? Don't ignore him, don't sell him short, because he may become a really big competitor before you have a chance to blink! It could be the next Drake-Williams Steel; so … like Satchel Paige famously said, "Don't look back … something might be gaining on you!"

Hugh W. Williams

The following was a contribution by Dave Williams, generous to Mike and Hugh, done for Ron Parks, who entered it in response to an Omaha Chamber of Commerce initiative to find stories about past "unsung" business leaders of this area:

November 1995

Hugh W. Williams and Michael C. Williams

Hugh and Mike Williams truly represent the spirit of Omaha business: dedicated, hard-working, tenacious, progressive, modest, and community-minded. They were instrumental in continuing one of Omaha's oldest family-held businesses, Drake-Williams Steel, Inc. As one of three teams of brothers spanning four generations, Hugh and Mike provided the leadership and determination necessary to guide the company through difficult times to prominence as one of the Midwest's foremost suppliers of fabricated steel.

Born, raised and educated in Omaha, Hugh and Mike both entered the business in the mid-50's. They inherited a legacy that began back in 1892 when their grandfather, also named

Hugh, joined Wilson & Drake Boiler Works, which at the time was little more than a blacksmith shop. Thanks in part to a gift of $3,000 from Andrew Carnegie, whom Hugh had met and impressed on a journey back to his native Wales, he purchased an interest in the company. In 1902, the company relocated to its present location at 23rd and Hickory Streets and was incorporated for the first time as the Drake-Williams-Mount Company.

Hugh and Mike's father Arthur, who acquired the nickname "Buckets" as a result of one of the company's key products, the drag-line bucket, teamed with his brother Walter during the second generation of Williams family ownership. The firm had major ups and downs during their tenure.

Drake-Williams-Mount Co. had serious problems to overcome when Hugh joined the company in 1956 and Mike shortly thereafter. It needed virtually complete re-equipping and a new direction. No new building or major addition had been accomplished since 1924 and the company owned no up-to-date equipment. The Company's business had been primarily the manufacture and repair of boilers, steel tanks, smokestacks and, since about 1910, crane scoops and drag-line buckets for excavation. With the end of World War II, the production demands required for the war effort had subsided. New methods of excavation were making the drag-line bucket obsolete.

Hugh and Mike insisted that the only way to save the company and the jobs it provided was to invest in more modern and efficient equipment, move the company into a new line of business, and to "pound the pavement" to search out new customers. It was clearly a case of young and aggressive management making the changes needed to save a company from bankruptcy while fighting an entrenched and conservative mindset.

The Company's focus was to shift to the steel warehousing and structural steel fabrication business. Less emphasis would be on the old-line business of drag-line buckets and boiler, tank and pressure vessel work. Drake-Williams-Mount Co. was changed to Drake-Williams Steel, Inc. in 1958 to reflect the change in the direction of the Company. In 1959, a bank loan was obtained to construct a new shop building addition and the purchase of some new and used equipment. Hugh, in charge of selling and estimating, found new customers and made sales. Mike handled the purchase of materials and also estimated jobs.

Hugh and Mike simply had to make the business a success; money had been borrowed from the bank for the Company's capital expenditures and loan obligations had to be met. As the Company struggled, two more events were to occur that would test the abilities and resolve of the two brothers.

Their father, "Buckets," retired in 1964 without any kind of plan in place regarding the transfer of ownership of the Company. In late 1965, part-owner and president Larry McMahon was killed in an auto accident. It was up to Hugh, at 34 years of age, and Mike, at the age of 27, to see the company through.

They were to overcome these obstacles and more. By 1969, sales had increased five-fold as compared to 1955. One of the most important events in the history of the company occurred in 1969 when adjacent land to the west of the plant was purchased so that vital plant expansion could take place. From 1969 to 1976, D-W-S built a new structural steel storage craneway and yard, a new structural fabrication shop and associated fabrication equipment, a warehouse, and finally a new office building.

The growth in the '60s and '70s was followed by a period of slower, but steady growth for Drake-Williams Steel. The company weathered a 3-year long recession in the early '80s which saw the doors of many steel fabrication companies close permanently throughout the nation and especially in the Midwest. In 1992, the land and buildings formerly operated by the Gate City Steel Corporation were purchased. Located on north 11th Street in Omaha, over 185,000 square feet are under roof on 13 acres of land. Hugh and Mike had the vision and foresight to see how this facility that had been the victim of years of decline could be renovated to meet Drake-Williams Steel's future expansion needs. In 1995, D-W-S met the requirements of LB775 and have added more than 30 jobs and an investment of more than $3 million since their application was made and approved in 1991.

In a "quiet" but vital industry, Hugh and Mike Williams have always avoided the recognition they deserve by crediting others, most notably their fine employees, for the firm's success. Under their progressive leadership, Drake-Williams Steel, Inc. quite literally provided the framework for many of Omaha's most prominent buildings. Its structural steel is found in the Kiewit Conference Center, W. Dale Clark Library, Children's Hospital, UNO's Fine Arts, Lab Science and HPER buildings, Creighton's Dental College, Law Center and Gymnasium, the Federal Reserve Building, and OPPD Energy Plaza.

Structural steel has been furnished for dozens of schools including Central, North, Westside, Benson and South High Schools. Steel for hundreds of Omaha business projects have been supplied to Kelloggs, Union Pacific Railroad, AT&T, ASARCO, Campbell Soup, Baker's Supermarkets, Mutual of Omaha, ConAgra, First Data Resources, AGP and many, many others.

After many years of work and community service together, the brothers decided to retire at the same time in January 1993. Hugh's two sons, Dave and John, have been with the firm since 1977, and currently serve as its President and CEO. Today, the company employs over 140 people. Drake-Williams Steel continues to grow as well as supply structural steel to help fuel Omaha's growth.

Dave Williams
Drake-Williams Steel, Inc.
2301 Hickory St.
Omaha, NE 68108
(402) 943-3422
FAX (402) 943-1570

This is our story, the history of Drake-Williams Steel, but I have a feeling that this is only the beginning, for much more is to come. A great future, a promising future, lies ahead, and I would surely fall short if I tried to predict what it holds in store. We have indicated the great strides in the tangibles—land, buildings, and equipment—all very important to facilitate growth and expansion. But, as we have seen throughout this story, the people are the chief ingredient, and the catalyst for real success. We have never been in better shape in this regard than at the present.

This history will have an epilogue, or maybe several in the coming years. These fine people who have been assembled under the D-W-S banner will be the ones to write the continuing story of this company of so many names. It has been my pleasure to take it this far and to chronicle the ups and downs and celebrate the success to date.

Perhaps somewhere my grandfather Hugh is smiling, and maybe even having a nice cool beer with old Teddy Evert. I hope we have managed to make them a little happiness.

SOURCES

A Dirty, Wicked Town: Tales of 19th Century Omaha by David L. Bristow, Caxton Press (2000)

Douglas County Historical Society, with thanks to Research Specialist Gary Rosenberg

Drake-Williams Steel archival materials (including photos)

A History of the City of Omaha and South Omaha by John Bell, Consul Butterfield and James Savage, Munsell and Company (1894)

History of the State of Nebraska Containing a Full Account of Its Growth from an Uninhabited Territory to a Wealthy and Important State, Western Historical Co., A.T. Andreas, Prop. (1882)

McAvoys City Directory, 1899, 1900 and 1904

Nebraska; An Illustrated History, Second Edition, Frederick C. Luebke

Nebraskans 1854 to 1904, The Bee Publishing Co. (1904)

Omaha: The Gate City and Douglas County Nebraska: A Record of Settlement, Organization, Progress and Achievement, S.J. Clarke Publishing Company (1917)

Omaha World-Herald archives

Our Times: The Illustrated History of the Twentieth Century, Turner Publishing, Atlanta (1995)

Personal letters, Georgia L. Getscher

Who's Who in Omaha, Robert M. Baldwin Publishing Co. (March 29, 1938)

INDEX

A

B

C

G

H

I

J

K

L

M

N

O

P

R

S

Y

Z

ABOUT THE AUTHOR

Hugh Williams was born in Omaha in 1931, where he has lived all his life. He is the son of Arthur W. and Janet (Willmarth) Williams. He has one older sister, Janet Fletcher of Orlando, Florida; and, of course, younger brother, Michael C. Williams, also of Omaha. He is a graduate of Central High School, attended the University of Nebraska, and is a graduate of the University of Nebraska at Omaha. He married Mary Jean Neely of Lincoln in 1953, and the couple has two sons, David W. and John A.

After two years service in the military, he was employed continuously at Drake-Williams Steel from 1956 until his retirement in 1992. He was president of the company from October 1965 until retirement in January 1992. Hugh and his brother, Mike, who served as vice president over that same period, began a process of modernization and expansion in 1965 that continues even to the present day. Landlocked at 23rd and Hickory streets, they first succeeded in acquiring several acres of ground to the west of the original plant, which gave them room to expand. Shops and an office building were constructed, but the growing business needed more room. An answer to this was the acquisition of part of the Omsteel facility, at 48th and Leavenworth. This was abandoned, however, several years later, when the Union Pacific Railroad ceased service along the belt-line route at that location.

Hugh and Mike became the second set of brothers to be in charge of the company, the first being their father, Arthur, and his brother, Walter. Now, the third set of brothers, David and John, manage affairs. In fact, Hugh and Mike remain envious of the spectacular growth and development of Drake-Williams under Dave and John! However, Hugh demands credit for the negotiation and purchase of the "new" plant at 11th and Seward streets in 1992. This is the plant belonging to Gate City Steel Co. for many years. Operations there under Hugh's two sons have been hugely successful.